Autorenkollektiv

Steric Effects in Drug Design

Introduction
M. Charton, I. Motoc

Features and Problems of Practical Drug Design
V. Austel

Topological Indices for Structure-Activity Correlations
A. T. Balaban, I. Motoc, D. Bonchev and O. Mekenyan

The Upsilon Steric Parameter — Definition and Determination
M. Charton

Molecular Shape Descriptors
I. Motoc

Volume and Bulk Parameters
M. Charton

Applications of Various Steric Constants to Quantitative Analysis of Structure-Activity Relationships
T. Fujita and H. Iwamura

AKADEMIE-VERLAG · BERLIN

Autorenkollektiv

Steric Effects in Drug Design

Autorenkollektiv

Steric Effects in Drug Design

With 40 Figures and 19 Tables

AKADEMIE-VERLAG · BERLIN
1983

Die Originalausgabe erscheint im Springer-Verlag
Berlin · Heidelberg · New York · Tokyo
in der Schriftenreihe „Topics in Current Chemistry",
Volume 114

Vertrieb ausschließlich für die DDR und die sozialistischen Länder

Erschienen im Akademie-Verlag, DDR-1086 Berlin, Leipziger Straße 3—4
Lizenznummer: 202 · 100/550/83
Printed in the German Democratic Republic
Gesamtherstellung: VEB Druckerei „Thomas Müntzer", 5820 Bad Langensalza
LSV 1275
Bestellnummer: 763 300 8 (6796)
07600

Managing Editor:

Dr. *Friedrich L. Boschke*
Springer-Verlag, Postfach 105280, D-6900 Heidelberg 1

Guest Editors of this volume:

Prof. *M. Charton*, Department of Chemistry, School of Liberal Arts and Sciences, Pratt Institute, The Clinton Hill Campus, Brooklyn, NY 11205, USA

Dr. *I. Motoc*, Max-Planck-Institut für Strahlenchemie, Stiftstr. 34–36, 4330 Mülheim a. d. Ruhr 1, FRG

Table of Contents

Introduction

Marvin Charton[1] and Ioan Motoc[2]

1 Department of Chemistry, School of Liberal Arts and Sciences, Pratt Institute, The Clinton Hill Campus, Brooklyn, NY 11205, USA
2 Max-Planck-Institut für Strahlenchemie, Stiftstraße 34–36, 4330 Mülheim a. d. Ruhr, FRG

Table of Contents

1 Introduction

Modern approaches to the design of bioactive molecules such as drugs, insecticides, herbicides, and fungicides is based on the quantification of bioactivity as a function of molecular structure. The seeds of this concept lie in the work of Meyer [1] and of Overton [2] who successfully demonstrated a dependence of bioactivity on a physicochemical parameter, the partition coefficient, which is a function of molecular structure. Vital to the development of the field was the concept of the receptor site, heralded by the work of Longley [3], stated and developed in depth by Ehrlich [4]. Biological activity according to this model depends on the recognition of a bioactive substrate (bas) by a receptor site, followed by binding of the bas to the receptor site. Realization of the dependence of bioactivity on configurations [5] led to the recognition that steric effects of one kind or another were a major factor in the potency of a bas. The work of Meyer and Overton was developed further particularly by Ferguson [6], whose work is fully described by Albert [7].

In a parallel development, structural effects on the chemical reactivity and physical properties of organic compounds were modelled quantitatively by the Hammett equation [8]. The topic is well reviewed by Shorter [9]. Hansen [10] attempted to apply the Hammett equation to biological activities, while Zahradnik [11] suggested an analogous equation applicable to biological activities. The major step forward is due to the work of Hansch and Fujita [12], who showed that a correlation equation which accounted for both electrical and hydrophobic effects could successfully model bioactivities. In later work, steric parameters were included [13].

Our objective in this work is to present surveys of the methods now available for the quantitative treatment of steric effects in the design of bioactive molecules. Commonly, this consists in the modification of a lead compound by structural changes which result in a set of bioactive substances. The bioactivity is determined and then related to structure. This is generally carried out by means of multiple linear regression analysis using a correlation equation of the type

$$Q_{ba,X} = T_1\tau_X + T_2\tau_X^2 + L\sigma_{IX} + D\sigma_{DX} + S\zeta_X + B_0 \qquad (1)$$

τ is a transport parameter,
σ_I and σ_D are the localized (field and/or inductive) and delocalized (resonance) electrical effect parameters, and
ζ is a steric parameter.

The correlation of bioactivity data with Eq. 1 or some relationship derived from it results, if successful, in a correlation equation called a quantitative structure activity relationship (QSAR).

When more than one substituent is present, the substituent effects may be assumed to be additive, or alternatively, each substituent may be parameterized separately.

Generally, the *transport parameter* used is the logarithm of the partition coefficient of the bas (bioactive substance) or some quantity derived from it. The partition coefficient is almost always determined between water and 1-octanol. Parameters obtained by chromatographic methods are being used with increasing frequency however. The term in τ^2 is introduced to account for the frequently observed parabolic dependence of a data set of bas on the transport parameter. Models other than parabolic have also been used, in particular, the bilinear model of Kubinyi [14]. In general, this behavior is accounted for by the fact that when a bas crosses a biomembrane it must first transfer from the aqueous phase to the biomembrane, and must then transfer from the other side of the biomembrane back into an aqueous phase.

Frequently, the *electrical effect parameters* are combined into a composite substituent constant. The well-known Hammett σ_m and σ_p constants [8,9] are examples of such composite electrical effect parameters. It is the steric parameter with which we are particularly concerned here.

In order to understand the *nature of steric effects* in molecular bioactivity it is necessary to have a model of the path by which a bas exerts its effect. Such a model has been proposed by MacFarland [15]. It involves the following steps:

1. The bas enters the organism and moves to a receptor site. In the course of its trip the bas will cross one or more membranes. This may be termed the transport step.
2. The bas is recognized by the receptor (rep) and a bas-rep complex is formed. Generally, this complex is held together by intermolecular forces and its formation is reversible. This may be called the complex formation step.
3. The bas-rep complex may undergo chemical reaction resulting in the formation and/or cleavage of covalent bonds. This step may be termed the chemical reaction step.

The structure of a bas may be written in the form XGY where X is the variable substituent (or substituents), Y is the active site at which bond formation or cleavage takes place, and G is the skeletal group to which they are attached. If no reaction occurs, then the bas has the form XG.

In the *transport step* the bas must be transferred from an aqueous phase to the membrane which may be represented as a lipid phase. The transport step can be considered to be a function of the difference in intermolecular forces between water and bas and those between lipid and bas. For a neutral bas these intermolecular forces include some combination of the following:

hydrogen bonding	(hb)
dipole—dipole	(dd)
induced dipole—induced dipole	(ii)
dipole—induced dipole	(di)
charge transfer	(ct)

For an ionic bas we may add to the list above:

ion—dipole	(Id)
ion—induced dipole	(Ii)

Of this list, hb interactions can be sensitive to steric effects. This has been established for amino acid transport parameters such as the hydrophobicities and partition coefficients. There is some evidence that charge transfer interactions are also subject to steric effects.

Of the remaining interactions, those which involve the dipole moment of the bas may be subject to an indirect steric effect as the preferred conformation of the bas may depend on steric effects and the dipole moment will depend on the conformation.

The Id and Ii interactions may be subject to steric effects resulting from steric hindrance to the solvation of the group which ionizes with the nature of the substituent.

Only the ii interaction is independent of steric effects.

In the *second step* the bas is recognized by the receptor site and the bas-rep complex forms. As was noted above, the complex is generally bonded by intermolecular forces. The bas is transferred from an aqueous phase to the receptor site. The receptor site is very much more hydrophobic than is the aqueous phase. It follows, then, that complex formation depends on the difference in intermolecular forces between the bas-aqueous phase and the bas-receptor site. The importance of a good fit between bas and receptor site has been known for many years. The configuration and conformation of the bas can be of enormous importance. Also important is the nature of the receptor. If the receptor is a cleft, as is the case in some enzymes, steric effects may be maximal as it may not be possible for a substituent to relieve steric strain by rotating into a more favorable conformation. In such a system, more than one steric parameter will very likely be required in order to account for steric effects in different directions. Alternatively, the receptor may resemble a bowl, or a shallow, fairly flat-bottomed dish. Conceivably it may also be a mound. In a bowl or dish, steric effects are likely to be very different from those in a cleft. Possible examples are shown in Fig. 1, 2, and 3.

In the *third step*, when it occurs, chemical bonds are made and/or broken. Steric effects in this case should resemble those generally observed for chemical reactivity

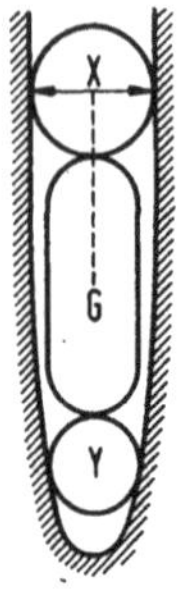

Fig. 1. "Cleft" receptor site (stylized). Steric effect of the substituent X is normal to the group axis, XG, and is shown by the arrows

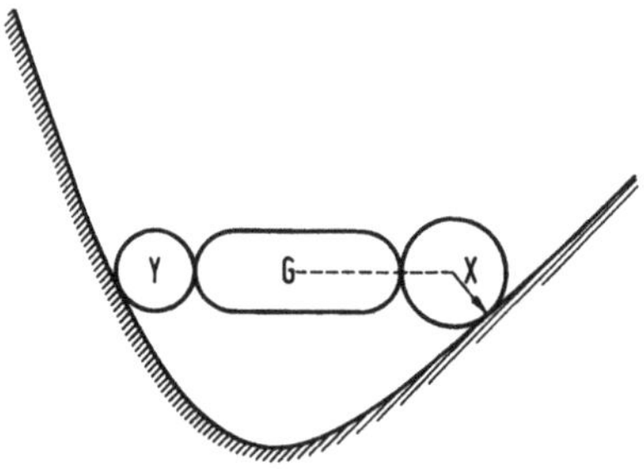

Fig. 2. "Bowl" receptor site

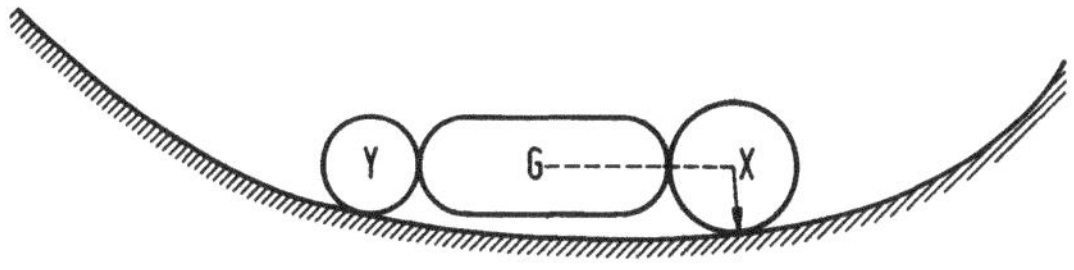

Fig. 3. "Dish" receptor site

if the substituent is in close proximity to the active site. Substituents distant from the active site may still exert a steric effect on the fit with the receptor site.

There are several distinct methods available for the parameterization of steric effects. The first group of parameters is defined from chemical reactivities (Taft E_S parameters and their modifications [16]), from Van der Waals radii and molecular geometries (Verloop [17]), and from a combination of these sources (Charton [18]). An alternative to these parameters is the use of the topological methods such as DARC-PELCO (Dubois and coworkers [19]), the branching equation (Charton [20]), molecular connectivity (Kier and Hall [21]) and minimal steric difference (Simon and Szabaday [22]). Steric parameters may also be obtained from force field calculations.

A second major area of interest in the design of bioactive molecules is that of the process of molecular recognition which is vital to the formation of the bas-rep complex. The functional groups required for recognition and activity, in their appropriate arrangement in space, or conformation, constitute the pharmacophore of a bas. The study of the possible conformations of a bas may be carried out by molecular mechanics or quantum chemical calculations. Another important factor is the nature and geometry of the receptor site.

Finally, it is necessary to consider the significance of "bulk" parameters such as molar volume, parachor and related quantities which have frequently been suggested as a measure of steric effects.

In the reviews which follow this introduction, we shall present a variety of methods for the treatment of steric effects.

2 Acknowledgements

We acknowledge with thanks the cooperation and support we have received from Springer Verlag in the course of this work. We would especially like to thank Dr. F. L. Boschke for his aid and guidance which were essential to our success.

3 References

1. Meyer, H.: Arch. Exptl. Pathol. Pharmakol. *42*, 110 (1899)
2. Overton, E.: Studien über die Narkose, Fischer, Jena 1901
3. Langley, J. N.: J. Physiol (London) *1*, 339 (1978)
4. Ehrlich, P.: Ber. dtsch. chem. Ges. *42*, 17 (1909)
5. Cushny, A. R.: Biological Relations of Optically Isomeric Substances, Balliere, Tindall and Cox, London 1926
6. Ferguson, J.: Proc. Royal Soc. *127B*, 387 (1939)
7. Alpert, A.: Selective Toxicity, Chapter 13, Methuan, London 1965

8. Hammett, L. P.: J. Am. Chem. Soc. *59*, 96 (1937)
9. Shorter, J.: Correlation Analysis in Organic Chemistry, Clarendon Press, London 1973
10. Hansen, O. R.: Acta Chem. Scand. *16*, 1593 (1962)
11. Zahradnik, R.: Arch. Int. Pharmadyn. Ther. *125*, 311 (1962)
12. Hansch, C., Fujita, T.: J. Am. Chem. Soc. *86*, 1616 (1964)
13. Hansch, C., Deutsch, E. W., Smith, R. N.: ibid. *87*, 2738 (1965)
14. Kubinyi, H.: Prog. Drug Res. *23*, 98 (1979)
15. McFarland, J. W.: ibid. *15*, 123 (1971)
16. a) Taft, R. W.: The Separation of Polar, Steric and Resonance Effects in Reactivity in: Steric Effects in Organic Chemistry (Newman, M. S. ed.), p. 556, Wiley, New York 1056
16. b) Shorter, J.: The Separation of Polar, Steric and Resonance Effects by the use of Linear Free Energy Relationships, in: Advances in Linear Free Energy Relationships (Chapman, N. B., Shorter, J. eds.), Plenum, New York 1972
17. Verloop, A., Hoogenstraaten, W., Tipker, J.: Drug Design *7*, 165 (1976)
18. Charton, M.: The Prediction of Chemical Lability Through Substituent Effects, in: Design of Biopharmaceutical Properties Through Prodrugs and Analogs (Roche, E. B. ed.), Am. Pharm. Assoc. Acad. Pharm. Sci., Washington, D.C. 1977
19. Dubois, J. E., Laurent, D., Aranda, A.: J. Chim. Phys. *70*, 1616 (1973)
20. Charton, M.: J. Org. Chem. *43*, 3995 (1978)
21. Kier, L. B., Hall, L. H.: Molecular Connectivity in Chemistry and Drug Research, Academic Press, New York 1976
22. Simon, Z., Szabaday, Z.: Studia Biophys. (Berlin) *39*, 123 (1973)

Features and Problems of Practical Drug Design

Volkhard Austel

Dr. Karl Thomae GmbH, Biberach an der Riss

Table of Contents

1 Introduction

The search for new drugs is one of the main objectives of medicinal chemistry. The biological properties of these drugs must be significantly superior to those of known drugs so as to improve therapeutical treatment of diseases. The desirable improvements are specified by the medicinal objective and comprise pharmacodynamic, e.g. type of effect, potency, selectivity, as well as pharmacokinetic properties, such as absorption, distribution and metabolic behaviour. These properties constitute the activity profile of a drug. Since there are no generally valid rules which relate chemical structure to certain activity profiles, new drugs can only be searched for in an empirical manner. Considering the complexity of the matter it is not surprising, therefore, that nowadys on an average between 6000 and 10000 compounds have to be synthesized and tested before one of them can be introduced into the market as a new drug. The demands on the biological properties of new drugs are increasing as are those on the corresponding experimental and clinical investigations. This does not only reduce the rate of success in improving medicamental therapy but also raises the costs of new developments considerably. Since the funds for drug research are limited one has to find means of reducing the rise in these costs. The medicinal chemist, like any other scientist involved in drug research, can contribute to this aim by using experimental capacity as economically as possible.

2 Drug Design in Practice

2.1 Economization of Empirical Drug Design

How can one proceed economically in fields of research which can only be explored empirically? This is possible by designing experiments so that a maximum of new information is obtained, i.e. by optimizing the information-expense ratio. For the medicinal chemist, the term "experiment" signifies in this context test compounds and the information refers to structure-activity information. At this point two main problems arise, i.e.

— how can structure-activity information be quantified?
— how can the amount of structure-activity information which a certain compound will give be determined prior to synthesis and biological testing?

Strictly speaking, there is no objective way of quantifying structure-activity information. At best one can subjectively estimate the amount of structure-activity information a group of compounds will give relative to another one. What criteria can such an estimate be based on? It is a generally accepted rule that similar compounds will show similar biological properties and hence also give similar structure-activity information. Assuming that in an idealized case every one out of a group of potential test compounds requires the same experimental expense for synthesis and testing and that two of these compounds are to be selected, we can envisage two extreme cases, i.e. one in which the two compounds are structurally very similar and alternatively, one in which two compounds do not very much resemble each other.

In the first case both compounds would give roughly the same structure-activity information, whereas in the second case largely independent information will be obtained. Therefore, the first case exemplifies a comparatively unfortunate choice since the information-expense ratio is only half as good as that of the second case. This very simplified example shows one of the main ways in which the medicinal chemist can influence the costs of drug research, i.e. through proper design and selection of test compounds. It is an essential even though not sufficient requirement of economical drug research that test compounds are chose so as to be mutually dissimilar.

2.2 Selection of Test Compounds

2.2.1 Similarity of Chemical Compounds

When are chemical compounds similar and how can similarity be objectively determined?

The similarity measure intuitively used by organic chemists is the number of structural features and their mutual arrangement which two compounds have in common. From this point of view adrenalin (*1*) and isoproterenol (*2*) would look rather similar and in fact both have similar biological properties as they are both agonists of β-adrenergic receptors. Both compounds are comparatively dissimilar to propranolol (*3*) even though a certain resemblance is still maintained. This situation is again reflected in the biological properties. Thus, propranolol can still interact with β-receptors but can no longer stimulate them, i.e. contrary to the former two compounds propranolol is an antagonist.

On the other hand toluene (*4*) and 5-methoxy-indole (*5*) would not be considered very similar from a conventional point of view, yet their hemolytic activity is the same (log 1/C = 1.93, where C is the minimal molar concentration which causes 100% hemolysis in rabbit erythrocytes [1]). Obviously, topological similarity or dissimilarily is not always a factor which is relevant for biological activity. In the present example the relevant feature is a physicochemical property of the compounds, i.e. lipophilicity as expressed by the octanol/water partition coefficient

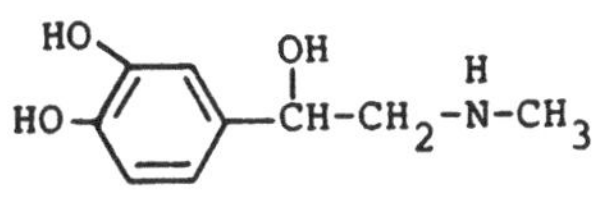

1

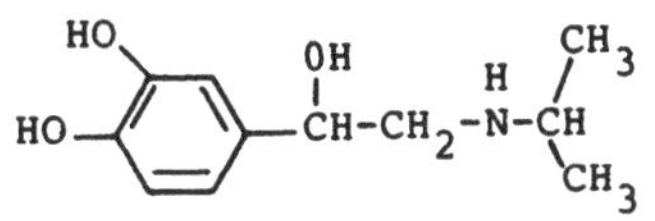

2

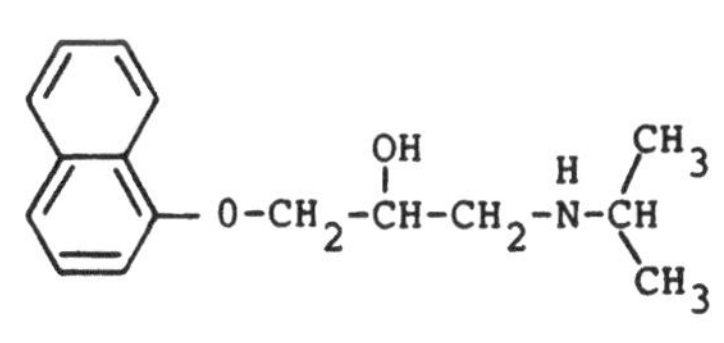

3

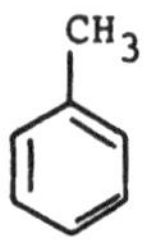

4

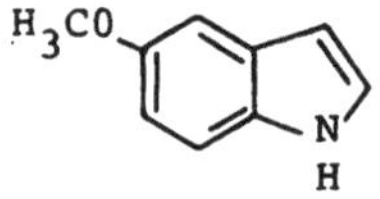

5

(log $P_{h/w}$ = 2.11 and 2.10 for *4* and *5* respectively)[2]. The idea that the biological properties of chemical compounds could be a function of their physico-chemical properties was proposed in the last century by Crum-Brown and Frazer[3,4]. Decisive progress in the understanding of the relationships between chemical structures of compounds and their biological activity came from the work of Hansch. In 1963, Hansch[5] showed that biological properties can be quantitatively related to physico-chemical properties of compounds with the aid of an extrathermodynamic model of drug action. Since then thousands of quantitative structure-activity-relationships have been found. Lists of several hundred examples have been published[6-8]. Considering this large background of evidence, there can be little doubt nowadays that the biological properties of chemical compounds are determined exclusively by their physico-chemical properties. For quantitative description one uses parameters which represent the corresponding property either explicitly or implicitly.

Thus, the commonly applied parameters π, σ and E_s are explicit representations of lipophilic, electronic and steric properties respectively. Indicator variables, on the contrary, frequently refer to fixed combinations of physico-chemical properties. For example, an indicator variable which denotes the presence of a 4-methoxy-group in general structure *6* refers simultaneously to all physico-chemical properties of this substituent, i.e. to its contribution to the lipophilic (π), electronic (σ), and steric properties (E_s) of the system.

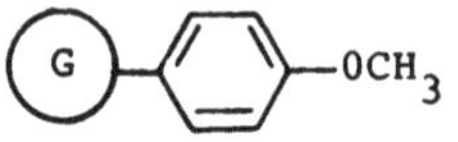

6

G is some basic pharmacophoric system

The Free-Wilson method of deriving quantitative structure-activity-relationships[10] uses implicit representations of physico-chemical properties and there are also numerous examples where indicator variables have been successfully included in the Hansch approach.

The possibility of describing chemical structures numerically with the aid of physico-chemical parameters and indicator variables puts us in the position to determine similarity or dissimilarity of chemical compounds more objectively. Chemical compounds can be represented as points in an n-dimensional space whose coordinates are formed by the parameters which are used to characterize the compounds. This space is therefore called parameter space. The distance of two points in parameter space can be taken as a measure of the similarity of the corresponding compounds [11]. The longer the distance the less similar the two compounds are.

As has been mentioned previously, mutual dissimilarily of the test compounds is a necessary but not sufficient criterion for a properly designed test series. In addition, the parameters which have been chosen to characterize the compounds must vary independently of oneanother. In order to fulfil both criteria an appropriate test series has to be designed so that the corresponding points are evenly distributed over the respective parameter space.

2.2.2 Determination of the Structural Area to be Investigated

Normally it is hardly feasible nor necessary to explore a certain parameter space to the full theoretically possible extent. Rather certain areas of parameter space are examined. These areas can be determined by intuition, synthetic accessibility or by structure-activity-relationships and structure-activity-hypotheses. Consider for example structure *6* to be a lead compound which needs to be optimized with respect to an activity profile consisting of two components, i.e. potency and oral effectivity (gastro-intestinal absorption). Let us for simplicity assume that we want to confine ourselves to variations of the substituent in 4-position (*7*) and that we

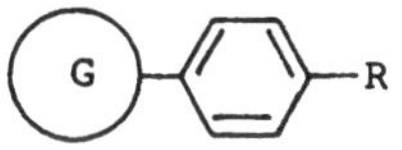

7

know size, lipophilicity and electronic properties to be the decisive factors for the biological properties under consideration. If we know for instance that a substituent whose MR-value (taken as a measure for size) exceeds 20 renders the respective compounds more or less inactive and that gastro-intestinal absorption is only significant if the lipophilicity contribution of R is larger than -0.5 (in π-terms), we can impose appropriate limitations on the area in parameter space which we want to investigate. With respect to electronic properties and to the maximum lipophilicity no limitations would exist a priori. The medicinal chemist will now by intuition and under accessibility considerations conceive a basic set of structures (in this simple case, substituents) which fall within the permissible area of parameter space, e.g. the ones shown in Table 1. In this Table lipophilicity, size and electronic effects are represented by π, MR and σ_p respectively. As can be seen from the values in Table 1, intuition and synthetic accessibility have brought about additional limitations to the investigated area in parameter space, i.e. -0.84 and 0.78 in σ_p-direction, and 1.98 as the upper limit in π-direction. In MR-direction the lowest possible value (0.92 for fluorine) is reached.

2.2.3 The Number of Test Compounds

Having defined a relevant parameter space or an area within it respectively and a basic set of possible structures we can now return to our original problem, i.e.

Table 1. Set of substituents which fall within a permissible area in a three dimensional parameter space formed by lipophilic (π), steric (MR) and electronic (σ_p) coordinates. The limitations are: MR < 20, $\pi > -0.50$ (values from Ref. [9])

No.	R	π	MR	σ_p
1	H	0.00	1.03	0.00
2	Br	0.86	8.88	0.23
3	Cl	0.71	6.03	0.23
4	F	0.14	0.92	0.06
5	NO_2	−0.28	7.36	0.78
6	CF_3	0.88	5.02	0.54
7	CH_3	0.56	5.65	−0.17
8	OCH_3	−0.02	7.87	−0.27
9	SCH_3	0.61	13.82	0.00
10	$NHCH_3$	−0.47	10.33	−0.84
11	$COOCH_3$	−0.01	12.87	0.45
12	C_2H_5	1.02	10.30	−0.15
13	OC_2H_5	0.38	12.47	−0.24
14	SC_2H_5	1.07	18.42	0.03
15	NHC_2H_5	0.08	14.98	−0.61
16	$N(CH_3)_2$	0.18	15.55	−0.83
17	$COOC_2H_5$	0.51	17.47	0.45
18	OC_3H_7	1.05	17.06	−0.25
19	$CH_2(NCH_3)_2$	−0.15	18.74	0.01
20	t-C_4H_9	1.98	19.62	−0.20

to select test compounds which are evenly distributed over a certain area in parameter space. The first decision which needs to be made refers to the number of test compounds. This number is mainly dependent on the dimensionality of the investigated parameter space and on the stage of the particular drug development project. In order to keep the variables which form the parameter space independent the number of test compounds must exceed the number of variables (dimensions of the parameter space) by at least one. This number will suffice in very early stages of lead optimization when a large number of potentially relevant parameters has to be considered. In more advanced stages when one is more confident about the relevant parameter space and when one tries to obtain a more detailed picture of the distribution of strongly and weakly active compounds in parameter space, two to four test compounds per parameter are a minimum requirement. Such numbers allow the location of the more interesting compounds in parameter space to be estimated. In very advanced stages when one tries to operate with quantitative structure-activity-relationships (QSAR) a minimum of six to ten compounds per dimension of parameter space are required in order to obtain meaningful results.

2.2.4 Methods for Selection of Test Compounds

After the number of test compounds which are to be selected from the basic set has been decided upon one can apply one of the series design techniques reported in the literature. Some of these methods apply to a restricted number of parameters (e.g. two in the Craig graphical method [12,13]) and structural variations (e.g. a limited number of aromatic and aliphatic substitution patterns as in the Topliss manual methods [14-16]). Others are, at least theoretically, not subject to such limitations but require computer facilities if more than two parameters are involved. Well known examples are the methods reported by Hansch et al. [17] and Wootton et al. [18]. The first method divides a basic set of compounds by means of cluster analysis, into several groups of mutually similar compounds. The number of groups corresponds to the number of samples (test compounds) which one wants to select. Out of every group one compound is chosen according to synthetic accessibility. The Wooton method selects test compounds by successive elimination of groups of similar compounds from the basic set. All the compounds which in parameter space lie within a certain distance D of a selected test compound form a group which is eliminated. Of the remaining compounds the one is chosen next which lies closest to the centre of gravity of the previously selected test compounds. D is chosen according to the desired number of test compounds. In our laboratory a method has recently been developed which allows test compounds to be selected manually in multidimensional parameter space [19]. This method makes use of 2^n factorial experimental design techniques. Thereby the level which a parameter adopts in a certain compound is specified as high (+), intermediate (0) or low (—). The test compounds are then chosen according to certain schemes, i.e. matrices of +, 0 and — signs in which the columns refer to parameters and the rows to compounds or respective substituents.

Table 2a. 2^2 factorial scheme for selection of test compounds in three-dimensional parameter space

compound no.	factor A	B	—AB
1	—	—	—
2	+	—	+
3	—	+	+
4	+	+	—
5	0	0	0

Table 2b. Selection of 5 substituents from Table 1 by the scheme of Table 2a

no.	substituent	factor π (A)	MR (B)	σ_p (—AB)
1	OCH_3	—0.02 (—)	7.87 (—)	—0.27 (—)
2	CF_3	0.88 (+)	5.02 (—)	0.54 (+)
3	$COOCH_3$	—0.01 (—)	12.87 (+)	0.45 (+)
4	OC_3H_7	1.05 (+)	17.06 (+)	—0.25 (—)
5	Br	0.86 (0)	8.88 (0)	0.23 (0)

An example is given in Table 2 where a 2^2 factorial scheme (with confounding) is applied in order to select five samples from the substituents shown in Table 1.

Table 2b reveals a fundamental problem of series design, i.e. that it is normally not possible to achieve an ideally even distribution of the samples over every area of every parameter space. Such difficulties arise for mainly three reasons, i.e.

1) structural modifications do not normally allow physicochemical properties and the corresponding parameters to be varied really continuously and completely independently of oneanother
2) synthetic accessibility makes it sometimes impractical to choose theoretically well suited compounds
3) certain areas in parameter space can for theoretical reasons not be represented by real structures.

An example of the latter case is the combination of small size and high lipophilicity. Thus, the second sample in Table 2 would have to fulfil such a condition. The CF_3-group which has been chosen for a corresponding representation must in reality be viewed as a compromise candidate whose lipophilicity is lower than ideal. Similar compromises appear in substituents no. 3 and 5 whose MR-values are somewhat too low. These compromises can be considered as examples of limitations due to synthetic accessibility (assuming that only the substituents referred to in the table can easily be introduced into the basic skeleton).

2.3 The Role of Structure-Activity-Relationships

After synthesis and biological investigation of the test compounds the medicinal chemist has two types of information in hand, i.e. one refering to the chemical structure and the respective physico-chemical properties and an other one which concerns the biological properties of the compounds. The way in which the latter properties depend on the former ones can be determined by structure-activity-analysis which yields structure-activity-relationships (detailed accounts can be found e.g. in refs. [20] and [21]). Structure-activity-relationships can belong to either one of three categories, i.e.

1) classifications based on topological features (qualitative structure-activity-relationships)
2) classifications based on explicitly or implicitly expressed physico-chemical properties (by application of numerical, mostly computer assisted, methods).
3) QSAR, (frequently derived by regression analysis based on physico-chemical models which refer to the respective biological property.)

In practical drug design, i.e. in lead optimization, only the first two categories play an important role. Thus, qualitative structure-activity-relationships are applied in the initial stage in order to determine the topological frame for optimization. Numerical classifications mainly serve to detect relevant physico-chemical properties and favourable and unfavourable areas in the corresponding parameter space. QSAR were initially thought to be very effective in predicting the "best" compound of a series. In practice, hardly any new drug has been found in this way. This experience is not surprising, since QSAR usually predict the compound which is "best" with regard to one component of the activity profile. This compound may however not be optimal or even acceptable with respect to the other components and may therefore not be the "best" one in view of the medicinal objective.

QSAR retain, however, a certain significance in late stages of lead optimization where they can serve to

1) check whether all relevant physico-chemical parameters have been sufficiently accounted for
2) elucidate the mechanism of action of a new drug.

2.4 Determination of the Relevant Parameters

Considering the last two paragraphs one encounters an other problem of systematic drug design: In order to select appropriate test compounds one needs to know the relevant parameters. These in turn can only be found through structure-activity-analysis based on the selected compounds. In other words, one needs to know the results which are obtained from the test series before this series can be designed properly. How can this problem be solved? As in other similar cases (e.g. the calculation of orbital energies in quantum mechanics) an iterative procedure can be applied. Such a procedure is visualized in Scheme 1.

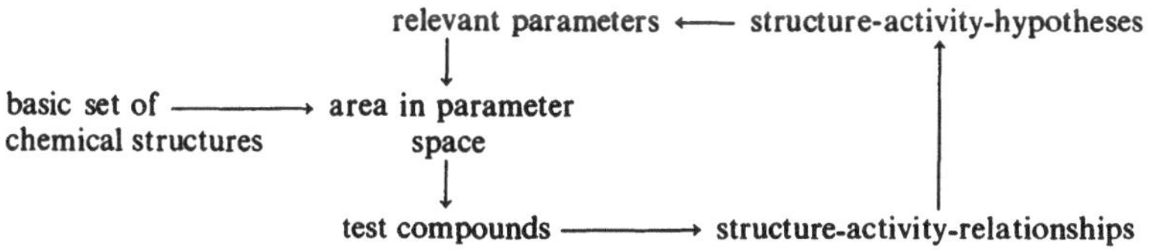

Scheme 1 Iterative determination of the relevant parameters

One starts with a structure-activity-hypothesis which refers to (hypothetically) relevant parameters. These parameters determine a parameter space in which a certain area is occupied by the basic set of structures from which the test compounds are to be chosen. After synthesis and testing of these compounds structure-activity-relationships are derived according to which the original structure-activity-hypothesis is modified. The modified structure-activity-hypothesis determines a modified parameter space and may allow interesting and less interesting parts of the corresponding area to be distinguished. Then, new test compounds are selected etc., until the derived structure-activity-relationships no longer differ significantly from the last modified structure-activity-hypothesis. In principle, this procedure has to be applied to every biological property which appears in the activity profile. By comparison of the corresponding structure-activity-relationships one can get an estimate of whether or not it is likely to find the desired new drug within the basic set of compounds [22].

2.5 Requirements for the Parameters

What requirements must parameters fulfil in order to be suitable in the search for new drugs? The type of parameter which is principally suitable depends on the respective problem, i.e. whether we are concerned with series design or with derivation of structure-activity-relationships. In the former case only parameters which can be calculated directly from the chemical structure can be used whereas in the latter case one can also (and preferably) apply measured parameters. Thus, measured log P values as well as calculated ones ($\Sigma\pi$) are suitable in structure-activity-analysis, whereas for series design lipophilicity can only be represented by the calculated values.

By using measured values in structure-activity-analysis more reliable structure-activity-relationships may be obtained, but higher experimental expense is normally required. When is this expense justified? This is only the case if the deviation between the calculated and the measured values of the physico-chemical property exceeds all the other errors which enter the respective structure-activity-relationships or if measuring requires less effort than calculation.

Which sources of error are involved in practical drug design? The most serious source of error is the relevance of the biological models for the therapeutic situation in humans. As a consequence, it is normally immaterial whether in lead optimization (as long as it is done in experimental animals) the "best" compound has been obtained. It is sufficient to find one (or a few) compound whose biological properties in the respective experiments are acceptable in view of the medicinal objective. Other significant errors stem from the fact that the "really" relevant physico-chemical properties will never be known completely. This is especially true for the initial stages of an optimization procedure, where the applied structure-activity-hypotheses are still poorly founded. A similar situation arises when many parameters would have to be used in order to quantify the potentially relevant physico-chemical properties. In order to keep the number of test compounds within a reasonable range one has to reduce the number of parameters by making simplifying assumptions. Thus in case of structure *8* we cannot generally expect R^1 and R^2 to

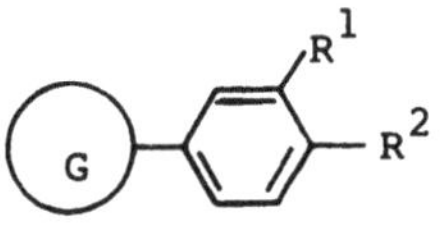

8

contribute equivalently to overall lipophilicity, i.e. we would need two separate parameters (π_1 and π_2) to express this property (eq. 1, where a_0 is a constant term, i.e. the contribution from the basic skeleton). In order to reduce the number of parameters one can now assume that $a_1 \approx a_2$ in which case lipophilicity can be quantified by one single parameter, i.e. $\Sigma\pi$ (eq. 2).

$$\log \quad = a_0 + a_1\pi_1 + a_2\pi_2 \qquad (1)$$

$$\log P = a_0 + A\Sigma\pi \qquad (2)$$

The need to limit the number of parameters becomes especially evident if molecular shape, which decisively influences the biological properties of chemical compounds, must be considered. Principally, shape can be precisely accounted for by the coordinates of all the atoms in the molecule. Even with rather small molecules (e.g. 20 atoms) one would need a prohibitive amount of parameters (e.g. 60) alone for representation of steric properties. Again, simplifying assumptions are made to reduce the number of parameters. Thus, one can for example assume that only the steric bulk in a certain position determines the biological properties, in which case a one-parameter representation may suffice, e.g. MR or van der Waals volume of the respective substituent.

In conclusion, for series design and largely also for evaluation of test series by structure-activity-analysis, the parameters which are used to describe physicochemical properties should meet the following condition:
They must be easily accessible for a large number of structural moieties. Examples are provided by lipophilicity constants (e.g. π, f) and to a lesser extent by electronic (e.g. various σ constants, R, F) and steric constants (e.g. E_s, υ) which have been listed for aromatic and aliphatic substituents [9,23]. These constants have been measured in model systems which need not necessarily be a good representation for the interaction of chemical compounds with biological structures. Therefore, application of these parameters in drug design involves a certain amount of error, which is, however, usually tolerable, especially in view of all the other sources of error mentioned above. This argument also applies in favour of parameters which are derived directly from the topology of molecules. Such parameters have the advantage of being available for every conceivable structural moiety. The price which one has to pay for this conveniance is usually a loss in accuracy with which the corresponding physicochemical property can be quantified. This draw-back is however not very significant as long as such parameters are applied in series design and in structure-activity-classifications. Their use in QSAR cannot generally be recommended.

An example for such a parameter is the steric constant S_b which can be used as an alternative to E_s or υ. This constant has been derived on the assumption that the steric effect of a substituent is mainly due to branching [24]. By regression

Table 3. Examples for S_b-values of some substituents

substituent	S_b
$\overset{\alpha}{H}$	0
$\overset{\alpha}{C}H_3$	1
$\overset{\alpha}{O}\overset{\beta}{C}H_3$	2
$\overset{\alpha}{N}(\overset{\beta}{C}H_3)_2$	3
$\overset{\alpha}{C}H_2-\overset{\beta}{C}H_2-\overset{\gamma}{C}H_3$	3
$\overset{\alpha}{C}H_2-\overset{\beta}{C}H_2-\overset{\gamma}{C}H_2-\overset{(\delta)}{C}H_3$	3

analysis we have found that only non-hydrogen atoms in α-, β- and γ-positions (and therefore branches from the α- and β-positions) of the substituent contribute significantly to its steric effect and that all these contributions are approximately equal [25]. Therefore, we can easily calculate a steric parameter (S_b) for every arbitrary substituent by simply counting the number of non-hydrogen atoms in α-, β- and γ-positions. Examples are given in Table 3. For ring structures and elements from the third, fourth and fifth rows of the periodic table simple corrections have to be introduced [25].

3 Summary and Outlook

Practical drug design is aimed at achieving certain activity profiles which render the corresponding compounds more valuable for therapeutic purposes than currently available drugs. Since the search for new drugs still proceeds mainly empirically, increasing demands on the quality of new drugs entail progressively increasing experimental expense. In order to prevent this expense from becoming prohibitive, empirical drug design needs to be economized. This can be done by careful selection of test compounds and by efficient extraction of structure-activity-information from the corresponding biological data. For both purposes one needs numerical representations of the physico-chemical properties of chemical compounds, since these properties are the determinants of biological activity. Physico-chemical properties can be more or less accurately described by respective parameters. At the present state of practical drug design it is very important that these parameters are easily accessible for a maximum number of structural moieties. The accuracy of representation is comparatively less important (at least for most practical purposes). This does not apply to mechanistic studies which are carried out in in-vitro models (enzyme and receptor binding studies). In such cases the biological data are rather exact and their analysis with respect to chemical structures or the corresponding physico-chemical properties requires the latter ones also to be described exactly.

Since mechanistic considerations and corresponding biological experiments are gaining increasing significance for practical drug design more exact physico-chemical descriptions of chemical structures will also be needed in this field. The conflict between accuracy and the number of required parameters will, however, persist as long as new drugs can only be searched for in predominantly empirical manner.

4 References

1. Rogers, S. K.: Proc. Soc. Exp. Biol. Med. *130*, 1140 (1969)
2. Hansch, C., Glave, W. R.: Mol. Pharmacol. *7*, 337 (1971)
3. Crum-Brown, A., Frazer: T. Trans. Roy. Soc. Edinburgh *25*, Pt. I, 151 (1868)
4. Crum-Brown, A., Frazer, T.: ibid. *25*, Pt. II, 693 (1869)
5. Hansch, C. et al.: J. Amer. Chem. Soc. *85*, 2817 (1963)

6. Hansch, C., Dunn, W. J.: J. Pharm. Sci. *61*, 1 (1972)
7. Hansch, C., Clayton, M. J.: J. Pharm. Sci. *62*, 1 (1973)
8. Seydel, J. K., Schaper, K.-J.: Chemische Struktur und biologische Aktivität von Wirkstoffen, p. 307ff. Weinheim—New York, Verlag Chemie 1979
9. ref. 8, p. 266ff.
10. Free, S. M., Wilson, J. W.: J. Med. Chem. *7*, 395 (1964)
11. Kowalski, B. R., Bender, C. F.: J. Amer. Chem. Soc. *94*, 5632 (1972)
12. Craig, P. N.: J. Med. Chem. *14*, 680 (1971)
13. Craig, P. N.: Adv. Chem. Ser. *114*, 115 (1972)
14. Topliss, J. G.: J. Med. Chem. *15*, 1006 (1972)
15. Topliss, J. G., Martin, Y. C.: in Drug Design (ed. E. J. Ariëns), Vol. V, p. 1, New York, Academic Press, 1975
16. Topliss, J. G.: J. Med. Chem. *20*, 463 (1977)
17. Hansch, C., Unger, S. H., Forsythe, A. B.: ibid. *16*, 1217 (1973)
18. Wootton, R. et al.: ibid. *18*, 607 (1975)
19. Austel, V.: Eur. J. Med. Chem. *17*, 9 (1982)
20. ref. 8, p. 115ff.
21. Martin, Y. C.: Quantitative Drug Design, New York—Basel, Marcel Dekker 1978
22. Austel, V., Kutter, E.: Arzneim.-Forsch. *31(I)*, 130 (1981)
23. Hansch, C., Leo, A.: Substituent Constants for Correlation Analysis in Chemistry and Biology, New York, J. Wiley, 1979
24. Charton, M.: J. Org. Chem. *43*, 3995 (1978)
25. Austel, V., Kutter, E., Kalbfleisch, W.: Arzneim.-Forsch. *29(I)*, 585 (1979)

Topological Indices for Structure-Activity Correlations

Alexandru T. Balaban[1], Ioan Motoc[2], Danail Bonchev[3] and Ovanes Mekenyan[3]

1 The Polytechnic, Organic Chemistry Department, Bucharest, Roumania
2 Max-Planck-Institut für Strahlenchemie, Mühlheim a. d. Ruhr, FRG
3 Higher School of Chemical Technology, Physical Chemistry Department, Burgas, Bulgaria

Table of Contents

This chapter deals with the description of the main topological indices and of related indicators for molecular constitution used in structure-activity relationships (QSAR). The topological indices are numerical quantities based on various invariants or characteristics of molecular graphs. For the convenience of the discussion, these indices are classified according to their logical derivation from topological invariants, rather than according to their chronological development.

Graph theory is largely applied to the characterization of chemical structures, as well as to structure-property and structure-activity correlations, by means of so-called topological indices. These are numerical quantities based on various invariants or characteristics of molecular graphs.

A graph **G** is a mathematical structure consisting of points (vertices) connected by lines (edges). Among the various kinds of graphs used in chemistry (chemical graphs), the molecular (constitutional) graphs attract much attention since constitutional formulas are depicted by points representing atoms, and by lines representing covalent bonds.

In the following, a brief description of the main topological indices and of related indicators for molecular constitution (such as polynomials, number sequences, codes, etc.) will be presented.

For the convenience of the discussion, these indices are classified according to their logical derivation from topological invariants, rather than according to their chronological development.

1 Topological Indices Based on the Adjacency Matrix

The basic mathematical structure which maps a certain molecule (molecular graph **G**) is the adjacency matrix of the molecular graph A(**G**). For a molecule having N atoms, A(G) is a square N × N matrix. Its entries a_{ij} have only two different values: 1, and 0, due to the fact that two atoms in a molecule are in binary relation, being either connected or not connected:

$$a_{ij} = 1, \text{ for i,j-adjacent}$$
$$a_{ij} = 0, \text{ otherwise}$$

As examples, the hydrogen-depleted graphs and the adjacency matrices of two test molecules, 2,3,4-tri-methylpentane ($\mathbf{G_1}$) and sec-butyl-cyclohexane ($\mathbf{G_2}$), are given below.

$$A(G_1) = \begin{array}{c} \\ 1\\2\\3\\4\\5\\6\\7\\8 \end{array} \begin{array}{c} \begin{array}{cccccccc} 1&2&3&4&5&6&7&8 \end{array} \\ \left[\begin{array}{cccccccc} 0&1&0&0&0&0&0&0\\ 1&0&1&0&0&1&0&0\\ 0&1&0&1&0&0&1&0\\ 0&0&1&0&1&0&0&1\\ 0&0&0&1&0&0&0&0\\ 0&1&0&0&0&0&0&0\\ 0&0&1&0&0&0&0&0\\ 0&0&0&1&0&0&0&0 \end{array}\right] \end{array}, \qquad A(G_2) = \begin{array}{c} \\ 1\\2\\3\\4\\5\\6\\7\\8\\9\\10 \end{array} \begin{array}{c} \begin{array}{cccccccccc} 1&2&3&4&5&6&7&8&9&10 \end{array} \\ \left[\begin{array}{cccccccccc} 0&1&0&0&0&1&0&1&0&0\\ 1&0&1&0&0&0&0&0&0&0\\ 0&1&0&1&0&0&0&0&0&0\\ 0&0&1&0&1&0&0&0&0&0\\ 0&0&0&1&0&1&0&0&0&0\\ 1&0&0&0&1&0&0&0&0&0\\ 1&0&0&0&0&0&0&1&0&1\\ 0&0&0&0&0&0&1&0&1&0\\ 0&0&0&0&0&0&0&1&0&0\\ 0&0&0&0&0&0&1&0&0&0 \end{array}\right] \end{array}$$

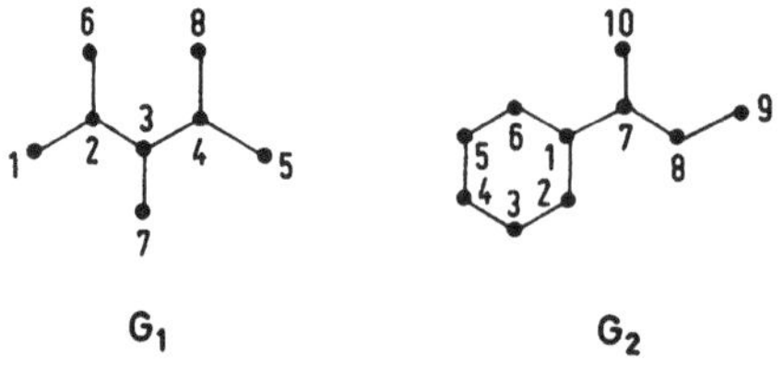

Fig. 1. Two illustrative graphs: 2,3,4-trimethylpentane (G_1) and sec-butyl-cyclohexane (G_2)

The adjacency matrix can function as a basis for the construction of several topological indices.

1.1 Total Adjacency Index

The simplest index is the sum A′ of all matrix elements:

$$A' = \sum_{i,j=1}^{N} a_{ij} \tag{1}$$

This index is redundant due to the symmetry of the adjacency matrix. It can be reduced to the sum A of the upper triangular adjacency submatrix, which is called [1] the *total adjacency of a molecule*:

$$A = \frac{1}{2} \sum_{i,j=1}^{N} a_{ij} = \frac{1}{2} A' \tag{2}$$

Owing to the fact that $a_{ij} = 1$ when there is a bond between atoms i and j, the total adjacency of a molecule equals the number of bonds in it. Burton [2] applied this quantity to cluster studies though not referring to its graph-theoretical origin.

For the two examples above, i.e., G_1 and G_2, one obtains A = 7 and 10, respectively. The total adjacency index, however, seems to be of little use for molecular studies since it can only distinguish between molecules having different *numbers of cycles* (i.e., cyclomatic numbers) μ:

$$A = N + \mu - 1 \tag{3}$$

Eqn. (3) is the known Euler equation connecting the number of vertices, edges and cycles in a planar graph.

1.2 The Zagreb Group Indices

The first numerical topological indices, based on the adjacency matrix, were introduced by the Zagreb group [3]:

$$M_1 = \sum_{i=1}^{N} v_i^2 \tag{4}$$

$$M_2 = \sum_{\text{all edges}} (v_i v_j) \tag{5}$$

where v_i is the degree of the vertex i in the hydrogen-suppressed graph. In mathematics, it is also called the valency of vertex i since v_i equals the number of bonds (edges) in the graph. Otherwise, v_i is the sum of all entries of i-th row in A(**G**):

$$v_i = \sum_{j=1}^{N} a_{ij}; \qquad A' = \sum_{i=1}^{N} v_i \tag{6}$$

v_i and v_j in eqn. (5) are the degrees of the two ends of the edge (ij).

Examples:

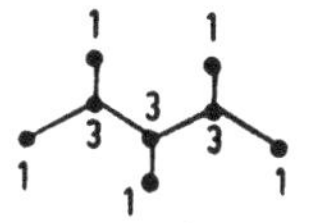

$$M_1 = 5 \times 1^2 + 3 \times 3^2 = 32$$

$$M_2 = 5 \times 1 \times 3 + 2 \times 3 \times 3 = 33, \text{ and}$$

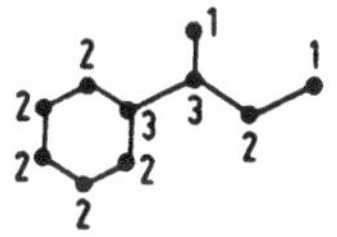

$$M_1 = 2 \times 1^2 + 6 \times 2^2 + 2 \times 3^2 = 44$$

$$M_2 = 1 \times 2 + 1 \times 3 + 4 \times 2 \times 2 + 3 \times 2 \times 3 + 3 \times 3 = 48$$

The two Zagreb indices are applicable to the calculation of π-electron energy of conjugated systems [4].

1.3 The Randić Connectivity Index

A connectivity index, χ_R, similar to the M_2 index was introduced by Randić [5] for characterization of molecular branching:

$$\chi_R = \sum_{\text{all edges}} (v_i v_j)^{-1/2} \tag{7}$$

For the graph G_1 one obtains:

$$\chi_R = 5(1 \times 3)^{-1/2} + 2(3 \times 3)^{-1/2} = 3.5536$$

and for G_2:

$$\chi_R = (1 \times 2)^{-1/2} + (1 \times 3)^{-1/2} + 4(2 \times 2)^{-1/2} + 3(2 \times 3)^{-1/2} + (3 \times 3)^{-1/2} = 4.8427$$

This index was applied to correlations with gas chromatographic retention index, boiling points, standard enthalpies of formation in gas phase, heats of solution, refractive indices, theoretically calculated total surface area of alkanes.

A generalized connectivity index was suggested by Randić, Kier and coworkers [6] considering the edge (ij) as the simplest specific case of a path of length h in the molecular graph, and extending the summation in Eq. (6) over all possible paths of length h:

$${}^{h}\chi_R = \sum_{\text{paths}} (v_i v_j \ldots v_{h+1})^{-1/2} \tag{8}$$

Here v_i, v_j, ..., v_{h+1} are the degrees of the vertices in the path of length h. The ${}^h\chi_R$ index may also be obtained [7] from the h-th power of the adjacency matrix, A^h.

The connectivity indices of order higher than three have not been used due to the expected small contribution to the molecular properties of the interactions between distant atoms. The calculation of ${}^2\chi_R$ for G_1 is presented below:

$${}^2\chi_R = 2\,(1\times3\times1)^{-1/2} + 6\,(1\times3\times3)^{-1/2} + (3\times3\times3)^{-1/2} = 3.34372$$

A further extension of this approach was done by Kier and Hall [8] so as to provide different values of the connectivity index for molecules depicted by one and the same graph, but differing by the chemical nature of atoms as well as by the presence of single, double or triple bonds. The valency of the atom i (vertex degree), v_i, is replaced by the atom connectivity Δ_i^v:

$$\Delta_i^v = Z_i^v - H_i \tag{9}$$

Z_i^v and H_i are the number of valence electrons of the atom i, and, respectively, the number of hydrogen atoms attached to this atom. The resulted connectivity index ${}^h\chi_R^v$ is given by:

$${}^h\chi_R^v = \sum_{\text{paths}} (\Delta_i^v \Delta_j^v \ldots \Delta_{h+1}^v)^{-1/2}$$

The generalized connectivity indices ${}^h\chi_R$ and ${}^h\chi_R^v$ have found numerous applications to structure-activity correlations, reviewed in the book of Kier and Hall [8]. Anaestetic activity was studied most detailed [6-13], along with nonspecific narcotic activity [8,14], enzyme inhibition [8,15,17] etc.

1.4 The Platt Index

(11)

Platt introduced [18] an index F which is calculated by summation of the number of bonds adjacent to each of the bonds in the molecule. In graph-theoretical terms, the Platt index is the sum of the degrees of all edges in the molecular graph:

$$F = \sum_{j=1}^{A} e_j$$

where the degree of the edge j, e_j, is the number of edges adjacent to edge j, and A is the total number of edges, already defined as the total adjacency of graph G. The F values for the graphs G_1 and G_2 are:

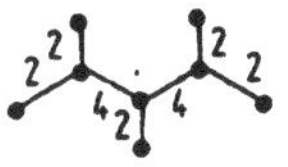

$$F = 5 \times 2 + 2 \times 4 = 18$$

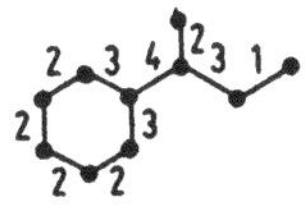

$$F = 1 \times 1 + 5 \times 2 + 3 \times 3 + 1 \times 4 = 24$$

The Platt index is defined by using the adjacency matrix of edges, exactly in the same way in which the index of vertex total adjacency, A′, was defined (i.e., eqns. 1 and 6). Hence, it could be called edge total adjacency. Otherwise, the indices A′ and F are called vertex and edge first neighbour sum. The Platt index was used [18] in correlations with some molecular properties in conjunction with other topological indices.

1.5 The Gordon-Scantlebury Index

This index, N_2, is defined [1] as the number of distinct ways in which an acyclic C—C—C fragment can be superimposed on the hydrogen-depleted molecular graph. The N_2 index expresses the number of all paths of length two, P_2, in the graph:

$$N_2 = \sum_i (P_2)_i \tag{12}$$

The N_2 values for the graphs G_1 and G_2 are:

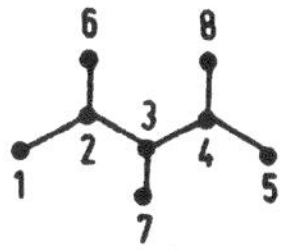

(P_2): 126, 123, 234, 237, 326, 345, 348, 437, 548
$N_2 = 9$

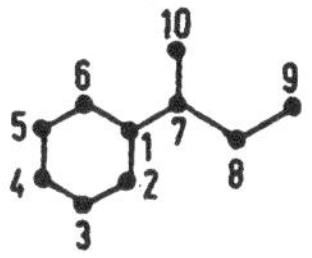

(P_2): 123, 234, 345, 456, 561, 612, 217, 617, 178, 1710, 789, 8710
$N_2 = 12$

The comparison with the F-values given above demonstrates the relation between the Gordon-Scantlebury and the Platt indices:

$$N_2 = \frac{1}{2} F \tag{13}$$

which can easily be proved for the general case.

Proceeding from eqn. (13) and the correspondence between A′ and F indices, one observes that the index N_2 corresponds to A. Hence, the Gordon-Scantlebury index has also the meaning of total edge adjacency of the graph.

Some simple relationships are also found among N_2 (or, F), the vertex degree, v_i, the index M_1, and the total adjacency A:

$$N_2 = \frac{1}{2} \sum_i v_i(v_i - 1) \tag{14}$$

$$N_2 = \frac{1}{2} M_1 - A$$

1.6 The Comparability Code and Index

Gutman and Randić [23] proposed an important concept for the comparability of chemical structures. They associated a numerical power sequence to a graph:

$$V = (v_1^{n_1} v_2^{n_2} \dots v_k^{n_k}) \tag{15}$$

V is called comparability (degree) code in which $v_1 = 1$, $v_2 = 2, \dots, v_k = k$, and $n_1, n_2, \dots, n_k$ are the respective frequency numbers (some of them may be zero). The code can be written more concise as an ordered sequence, omitting the vertex degrees since they are uniquely defined:

$$V = (n_1, n_2, \dots, n_k)$$

For the graphs G_1 and G_2 one obtains:

$$V = (1^5\ 2^0\ 3^3) = (5,\ 0,\ 3)\,, \quad \text{and, respectively,}$$
$$V = (1^2\ 2^6\ 3^2) = (2,\ 6,\ 2)\,.$$

The importance of this code lies the formulation of a criterion which classifies two structures as comparable, or non-comparable, making thus possible the ordering of structures. Muirhead's criterion [24] for comparability of functions is used: two sequences of numbers in non-ascending order, $(m_1, m_2, \dots, m_k)$ and $(m_1', m_2', \dots, m_k')$ where $\sum_{i=1}^{k} m_i = \sum_{i=1}^{k} m_i'$, can be compared and the precedence can be assigned to the first sequence if:

$$\sum_{i=1}^{p} m_i \geq \sum_{i=1}^{p} m_i'\,, \qquad 1 \leq p \leq k \tag{17}$$

Using eqn. (17), the V codes attached to the graphs G_1 and G_2 are transformed into Muirhead's sequences $V' = (m_1 = n_1, m_2 = n_1 + n_2, \dots, m_k = n_1 + n_2 + \dots + n_k)$ as:

$$V' = (5,\ 5+0,\ 5+0+3) = (5,\ 5,\ 8)\,, \quad \text{for the graph } G_1\text{, and}$$
$$V' = (2,\ 2+6,\ 2+6+2) = (2,\ 8,\ 10)\,, \quad \text{for the graph } G_2.$$

Because the conditions (17) do not hold, the two structures are non-comparable.

The concept of structure comparability is of practical use in the search of correlations between topological indices and molecular properties. It does not directly contain a numerical comparability index (through the authors [23] use this term for what we called above comparability code). Two such indices proposed here, as a modification of the M_1 index, and are denoted as M_3' and M_3'':

$$M_3' = \sum_{i=1}^{k} n_i^2, \qquad M_3'' = \sum_{i=1}^{k} m_i^2 \tag{18}$$

For the graph G_1 one obtains $M_3' = 34$, $M_3'' = 144$, and for G_2 the values are $M_3' = 44$, $M_3'' = 168$.

Another way for calculating comparability indices is possible on the basis of information theory.

1.7 The Smallest Binary Notation (SBN) of Randić

Randić transformed [25] the adjacency matrix of the graph into binary notation by permuting rows and columns until, on reading sequentially the rows, the smallest binary number resulted. Thus, a unique numbering of the graph vertices is provided. The resulted *smallest binary notation* (SBN) is supposed to solve the graph isomorphism problem, but it is to large to be applied as a normal topological index.

Illustratively, the smallest binary notation of the graph G_1 is:

SBN : 00000001 ¦ 00000001 ¦ 00000010 ¦ 00000010 ¦ 00000100 ¦ 00001011 ¦
00110100 ¦ 11000100

(The vertical dashed lines separate the contributions of the vertices 1 to 8 respectively), and the corresponding vertex numbering is shown below:

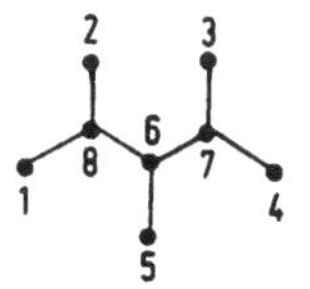

1.8 The Largest Eigenvalue Index

Lovasz and Pelikan [26] found that the largest eigenvalue, x_1, of the characteristic polynomial of the graph is a fairly sensitive measure of molecular branching. Hence, this eigenvalue was also examined as a topological index. For the graph G_1 it is $x_1 = 2.136$, while for the graph G_2 it is $x_1 = 2.214$. An explanation of the specific behaviour of x_1 was given by Cvetković and Gutman [27], who proved its connection with the number of walks in the graph.

2 Topological Indices Based on the Distance Matrix

The distance matrix D(G) of a graph G is another important graph-invariant. Its entries d_{ij}, called distances, are equal to the number of edges connecting the vertices i and j on the shortest path between them. Thus, all d_{ij} are integers, including $d_{ij} = 1$ for nearest neighbours, and, by definition, $d_{ii} = 0$. The distance matrix can be derived readily from the adjacency matrix:

$$D(G) = A(G) + \sum_{i=1}^{d_{max}} B_i(G) \tag{19}$$

Here $B_i(G)$ stands for matrices containing as single non-zero entries $b_2 = 2$, $b_3 = 3$ etc. For example, B_2 contains the shortest paths between the second neighbours, B_3 between the third neighbours etc.
The distance matrices of the graphs G_1 and G_2 are shown below:

$$D(G_1) = \begin{array}{c} \\ 1\\2\\3\\4\\5\\6\\7\\8 \end{array} \begin{array}{c} \begin{array}{cccccccc} 1&2&3&4&5&6&7&8 \end{array} \\ \left[\begin{array}{cccccccc} 0&1&2&3&4&2&3&4\\ 1&0&1&2&3&1&2&3\\ 2&1&0&1&2&2&1&2\\ 3&2&1&0&1&3&2&1\\ 4&3&2&1&0&4&3&2\\ 2&1&2&3&4&0&3&4\\ 3&2&1&2&3&3&0&3\\ 4&3&2&1&2&4&3&0 \end{array}\right] \end{array}, \quad D(G_2) = \begin{array}{c} \\ 1\\2\\3\\4\\5\\6\\7\\8\\9\\10 \end{array} \begin{array}{c} \begin{array}{cccccccccc} 1&2&3&4&5&6&7&8&9&10 \end{array} \\ \left[\begin{array}{cccccccccc} 0&1&2&3&2&1&1&2&3&2\\ 1&0&1&2&3&2&2&3&4&3\\ 2&1&0&1&2&3&3&4&5&4\\ 3&2&1&0&1&2&4&5&6&5\\ 2&3&2&1&0&1&3&4&5&4\\ 1&2&3&2&1&0&2&3&4&3\\ 1&2&3&4&3&2&0&1&2&1\\ 2&3&4&5&4&3&1&0&1&2\\ 3&4&5&6&5&4&2&1&0&3\\ 2&3&4&5&4&3&1&2&3&0 \end{array}\right] \end{array}$$

2.1 The Wiener Index

Approximately at the same time when Norbert Wiener wrote this fundamental work on cybernetics, Harold Wiener [28] proposed the first structural index of topological nature. In fact, H. Wiener defined the so-called *path number* w as the sum of the number of bonds separating all pairs of atoms in saturated hydrocarbons. The graph theoretical basis of this index was discussed by Hosoya who extended the definition to cyclic graphs. This extension, however, required to take as a basis the distances, and not all paths in the molecular graph (the two quantities coincide in the case of acyclic graphs). Recognizing the pioneering role of Wiener, the total number of distances between all pairs of vertices in acyclic and cyclic molecules is nowadays termed the Wiener index (or, w index). It is easily shown that this index equals the half-sum of the off-diagonal elements of the distance matrix:

$$w = \frac{1}{2} \sum_{i,j} d_{ij} \tag{20}$$

For the graphs G_1 and G_2 one obtains w = 65, and w = 121, respectively.
Rouvray[30,31] proposed as a topological index the sum R of all matrix elements of D(G):

$$R = \sum_{i,j} d_{ij} \tag{21}$$

Obviously, R = 2w, similarly to A' = 2A for the total adjacency. Proceeding from this analogy one can call either w or R the total distance of the graph, giving a preference to the Wiener index since, due to the symmetry relative to the main matrix diagonal, the Rouvray index is redundant. Applications of w = R/2 are discussed in the following section.

2.2 The Polarity Number

Wiener has also formulated[28] the so-called polarity number, p. This is the number of pairs of vertices, separated by three edges, or otherwise, half of the number of distances of lengths three, d_3, in D(G), or half of the sum of all entries in the B_3 matrix as:

$$p = \frac{1}{2} \sum_{i} d_{3,i} = \frac{1}{2} \sum_{i,j} b_{3,ij} \tag{22}$$

As seen from $D(G_1)$ and $D(G_2)$ above, p = 8 and, respectively, p = 11.
One can also see that the Gordon-Scantlebury N_2 index[19] can similarly be expressed as half the number of distances of length two, but for acyclic graphs only:

$$N_2 = \frac{1}{2} \sum_{i} d_{2,i} = \frac{1}{2} \sum_{i,j} b_{2,ij} \tag{23}$$

The Wiener index and polarity number have been applied by Wiener[28] and Platt[18] to correlations with boiling points, heats of formation and vaporization, molecular volume and molecular refraction of alkanes. High correlations were found for the Rouvray index with a number of physical properties of alkanes, alkenes, alkynes and arenes. The Wiener index was also correlated with the gas-chromatographic retention index[32-34]. This index was used as a basis for expressing the features of molecular branching[30,34,35], and cyclicity[36-39] in a number of structural rules which reflect the main molecular properties. The good results obtained by means of the Wiener index are related with the fact that this index has a minimum for the most compact structures out of an ensemble of isomeric molecules. This also prompted the application of the Wiener index to the modeling of crystal growth[40] and crystal vacancies[41,42], as well as to the prediction of the energy gaps[43,44] and physico-chemical properties of infinite conjugated polymers[45].

2.3 Distance Sum Index

This is another index, which was independently developed by Polansky[46] for expressing some graph properties, including the relation with the Wiener index for important classes of molecular graphs, and by Bonchev, Balaban and Mekenyan[47] as an additional criterion for the graph center determination (see also paragraph 3). The two groups of authors called this index "*the distance number*" and "*the distance rank*", respectively. We prefer to introduce a more systematic name, by analogy with the adjacency matrix. In the latter, the sum of all entries of the i-th row specifies the degree of the vertex i. Similarly, the sum of all entries of the i-th row in the distance matrix D defines the distance sums, $v_{D,i}$, of the vertex i:

$$v_{D,i} = \sum_{j=1}^{N} d_{ij} \tag{23'}$$

The values of these distance sums are given below at each vertex in graphs G_1 and G_2:

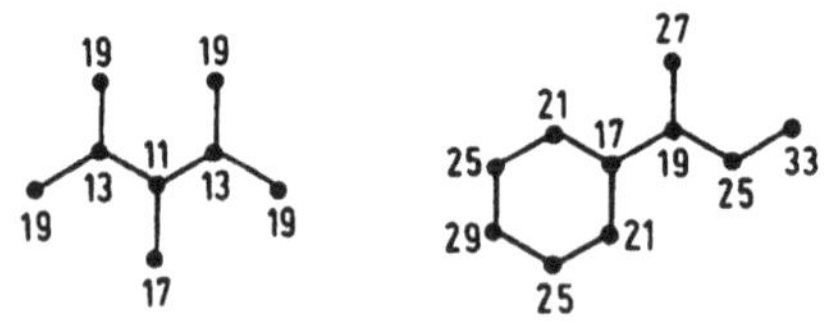

Summation for all vertices in the graph gives an index which for G_1 is 130, and for G_2 is 242.

2.4 Average Distance Sum Connectivity Index (Balaban)

Balaban applied [47a, b] a Randić-type formula (7) to distance sums instead of vertices and created thereby the most discriminating (least degenerate) topological index proposed so far:

$$J = \frac{q}{\mu + 1} \sum_{\text{adjacencies } i,j} (v_{D,i} v_{D,j})^{-1/2} \tag{24}$$

For simple graphs (saturated molecules) q is the number of edges. Averaged distance sums:

$$\bar{v}_{D,i} = v_{D,i}/q\,, \quad \text{and} \quad \bar{v}_{D,j} = v_{D,j}/q$$

were used in formula (7) instead of distance sums (resulting in the above formula for J) because distance sums increase in approximately parallel fashion with q for the same type of branching. The cyclomatic number μ was defined in formula (3) and was introduced in the formula of J because the presence of cycles markedly reduces the distance sums.

For multiple edges (unsaturated or aromatic systems) fractional distances are introduced in the distance matrix so that fractional distance sums result on summation: if the bond order between vertices i, j is b, then 1/b is the entry in the row/

column i/j. As examples, J = 2.4744 for G_1 and J = 2.2395 for G_2. For aromatic or unsaturated systems:

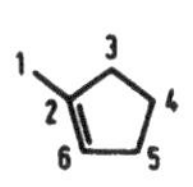

1

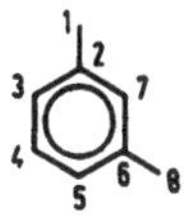

2

the D matrices are:

D(1) =

	1	2	3	4	5	6
1	0	1	2	3	2.5	1.5
2	1	0	1	2	1.5	0.5
3	2	1	0	1	2	1.5
4	3	2	1	0	1	2
5	2.5	1.5	2	1	0	1
6	1.5	0.5	1.5	2	1	0

D(2) =

	1	2	3	4	5	6	7	8
1	0	1	5/3	7/3	3	7/3	5/3	10/3
2	1	0	2/3	4/3	2	4/3	2/3	7/3
3	5/3	2/3	0	2/3	4/3	2	4/3	3
4	7/3	4/3	2/3	0	2/3	4/3	2	7/3
5	3	2	4/3	2/3	0	2/3	4/3	5/3
6	7/3	4/3	2	4/3	2/3	0	2/3	1
7	5/3	2/3	4/3	2	4/3	2/3	0	5/3
8	10/3	7/3	3	7/3	5/3	1	5/3	0

and the distance sums:

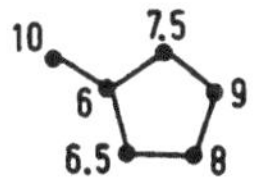

J = 2.8823

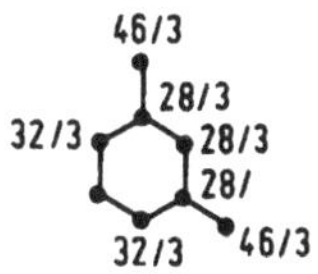

J = 3.0777

For all cycloalkanes having N = 2k carbon atoms, J = 2.0000 irrespective of k. It can be demonstrated that for linear alkanes J increases asymptotically with increasing N towards π; cycloalkanes with odd N tend asymptotically towards J = 2.0000; linear conjugated polyenes tend towards $4\pi/3$, and linear conjugated polyynes tend towards $2\pi/3$. On the other hand, with increasing branching, J increases indefinitely. Therefore, J reflects strongly the molecular branching and very little the molecular size (i.e., the value of N). For alkanes, J gives approximately the same ordering as I_D^W (see Sect. 4.3.). The only general degeneracy which was found till now for isomeric graphs is derived from the following monocyclic structures:

$$\overline{(CH_2-CRR')}_{2k} \text{ and } \overline{(CHR-CHR')}_{2k}$$

with integer k and R,R' ≠ H being alkyls (identical or different). Thus the following graph pairs have degenerate J values:

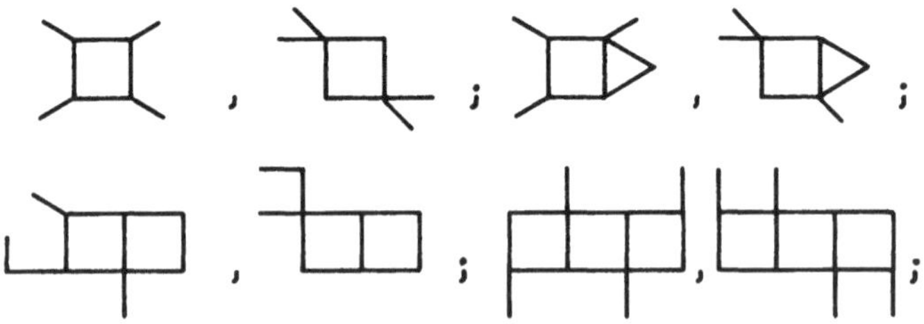

; etc.

The smallest two alkanes having degenerate J values have 22 vertices, where the total number of isomers (not counting stereoisomers) is 2,278,658. It will be seen, however, from Table 1 (see Sect. 4.7.) that the degeneracy for J is much lower than for any other single index.

Some of the following headings do not relate to indices in the proper sense (i.e., a single numerical value for one structure), but are included for convenience.

2.5 The Smolenskii Additivity Function

It is known that extensive molecular properties can be calculated within an additive scheme from the contributions of atoms, atomic groups, rings, and multiple bonds [48-50]. Smolenskii [51] formalized this method in graph-theoretical language, expressing additive properties of hydrocarbons by means of contributions of different subgraphs of the molecular graph **G**:

$$f(\mathbf{G}) = a_0 + \sum_{k=1}^{A} a_k X_k \tag{25}$$

where a_0 and a_k are experimentally-determined constants for the property under study, while X_k is a specified subgraph with k edges of the graph **G** having A edges. In a first-order approximation, the property f is expressed in terms of the number of three subgraphs: paths of lengths one, two and three, denoted by X_1, X_2 and X_3. In a better (second) approximation X_2 and X_3 are partitioned into subclasses according to the degree of the vertices in the hydrogen-suppressed graph **G** which are not endpoints of the path. The Smolenskii subgraphs for the graphs G_1 and G_2 are shown below:

Subgraphs (first approximation)	The graph G_1	The graph G_2
	$X_1 = 7$	$X_1 = 10$
	$X_2 = 9$	$X_2 = 12$
	$X_3 = 8$	$X_3 = 14$

(second approximation)		
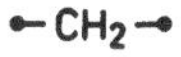	$X_2^1 = 0$	$X_2^1 = 6$
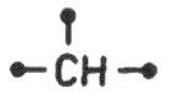	$X_2^2 = 9$	$X_2^2 = 6$
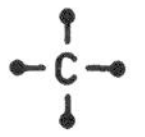	$X_2^3 = 0$	$X_2^3 = 0$
	$X_3^1 = 0$	$X_3^1 = 4$
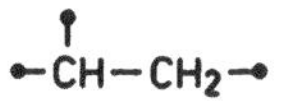	$X_3^2 = 0$	$X_3^2 = 6$
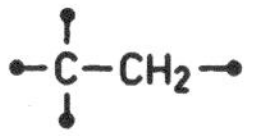	$X_3^3 = 0$	$X_3^3 = 0$
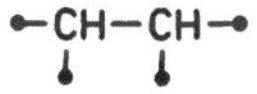	$X_3^4 = 8$	$X_3^4 = 4$
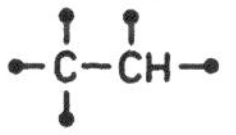	$X_3^5 = 0$	$X_3^5 = 0$
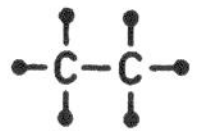	$X_3^6 = 0$	$X_3^6 = 0$

It is seen that the Smolenskii term X_2 is identical to the N_2 index, while the term X_3 is identical (in the case of acyclic graphs only) to the polarity number p.

2.6 The Altenburg Polynomial

The Wiener number can also be calculated by the formula

$$w = \sum_i d_i g_i \tag{26}$$

where g_i is half the frequency number of distance d_i in the distance matrix D(G).

Altenburg [52] transformed eqn. (26) into a polynomial $P_A(G, a)$ as:

$$P_A(G, a) = \sum_i a_i g_i \tag{27}$$

using the indexed variable a_i. The latter has no special significance, being introduced only for enumeration in polynomial form.

Thus, for the graphs G_1 and G_2 one obtains:

$$P_A(G_1, a) = 7a_1 + 9a_2 + 8a_3 + 4a_4$$
$$(W = 1\times 7 + 2\times 9 + 3\times 8 + 4\times 4 = 65)$$

$$P_A(G_2, a) = 10a_1 + 12a_2 + 11a_3 + 7a_4 + 4a_5 + a_6$$
$$(s = 1\times 10 + 2\times 12 + 3\times 11 + 4\times 7 + 5\times 4 + 6\times 1 = 121)$$

Hosoya [29)] extended the Altenburg polynomial (originally devised for acyclic graphs) to cyclic graphs.

Altenburg also introduced [52c)] the squared radii of alkanes:

$$R^2(G) = \frac{1}{N^2} \sum_i i g_i = \frac{W}{N^2}$$

and found linear correlations $R^2(G)$ *vs* χ_R.

2.7 Mean Square Distance Topological Index

On testing for various exponents k = 1, 2, 3 or 4 in formula

$$D^{(k)} = \left[\sum_{d_i=1}^{d_{i,\max}} g_i d_i^k \Big/ \sum_{d_i=1}^{d_{i,\max}} g_i \right]^{1/k}$$

it was found [47b)] that k = 2 afforded for alkanes the smallest degeneracy. For cyclic graphs, however, $D^{(2)}$ has a fairly high degeneracy.

For acyclic graphs only, if in the above formula only distances between endpoints are taken into account, the related endpoint mean square topological index $D_1^{(2)}$ results.

Both $D^{(2)}$ and $D_1^{(2)}$ decrease with increasing branching. Whereas $D^{(2)}$ induces in alkanes an ordering similar (but not identical) to that caused by $\bar{I}^E$, $D_1^{(2)}$ gives an ordering resembling that induced by $\bar{I}_D^W$.

Examples for graphs G_1 and G_2 are:

G_1:

d_i	1	2	3	4
g_i	7	9	8	4

$D^{(2)} = 2.528$, $D_1^{(2)} = 2.464$

G_2:

d_i	1	2	3	4	5	6
g_i	10	12	11	7	4	1

$D^{(2)} = 3.000$

2.8 Atomic and Molecular Distance Code and Index

Randić introduced [53] the so-called atomic and molecular path codes in an attempt to characterize uniquely molecular structures. The atomic (or vertex) path code of the atom j, VPC(j), represents the power series:

$$VPC(j) = p_1^{n_1} p_2^{n_2} \dots p_{max}^{n_m} \tag{28}$$

where p_i is the path of length i, and n_i is its frequency number.

The molecular (or graph) path code, GPC(G), is obtained similarly by summation over all atoms (vertices) in the molecule (graph **G**):

$$GPC(G) = p_1^{m_1} p_2^{m_2} \dots p_{max}^{m_m} \tag{29}$$

where $m_i = \frac{1}{2} \sum_{j=1}^{N} n_{ij}$, N being the total number of vertices, and $^1/_2$ stands for avoiding the double counting of each path.

More frequently, the two codes are used in linear form:

$$VPC'(j) \ = (n_1, n_2, \dots, n_m) \tag{28'}$$

$$GPC'(G) = (m_1, m_2, \dots, m_m) \tag{29'}$$

Though these codes failed to characterize uniquely the molecular structure [54], they are useful for ordering structures as basis for systematic searches for regularities in molecular data [55], fragment [56] and ring [57] search, for recognition of structural similarity [54] in molecules, for the prediction of chemical shift in NMR-data, etc.

For polycyclic structures it seems reasonable to introduce [59] vertex and graph distance codes, VDC and GDC, instead of path codes:

$$VDC(j) = d_1^{g'_1} d_2^{g'_2} \dots d_{max}^{g'_m} \tag{30}$$

$$GDC(G) = d_1^{g_1} d_2^{g_2} \dots d_{max}^{g_m} \tag{31}$$

where $g_i = \frac{1}{2} \sum_{j=1}^{N} g'_{ij}$ is the same as defined in Eq. (26), and g'_i is the frequency number of distance d_i for the vertex j. The linear form of these codes is obvious. Clearly, for acyclic graphs hold:

$$VDC(j) \ = VPC(j) \tag{32}$$

$$GDC(G) = GPC(G) \tag{33}$$

The above codes can be converted into topological indices [59] as:

$$\mathrm{VDI} = \sum_i (g_i')^2 \quad (34); \qquad \mathrm{GDI} = \sum_i (g_i)^2 \quad (36)$$

$$\mathrm{VPI} = \sum_i (n_i)^2 \quad (35); \qquad \mathrm{GPI} = \sum_i (m_i)^2 \quad (37)$$

Other details concerning the above indices may be found in Refs. [53–58].

2.9 The Hosoya Index

An important topological index was introduced by Hosoya [29,60,61]:

$$Z = \sum_{k=0}^{[N/2]} p(G, k) \tag{38}$$

where $p(G, k)$ is the number of ways in which k edges are chosen from the graph **G** so that no two of them are adjacent; N/2 in the Gauss square brackets is the smallest integer not exceeding the real number in them. By definition, $p(G, 0) = 1$, while $p(G, 1)$ equals the number A of edges in the graph. An illustration for graph $\mathbf{G}_1$ is presented below ($p(G_1, 0) = 1$, $p(G_1, 1) = 7$).

$p(G_1, 2) = R$:

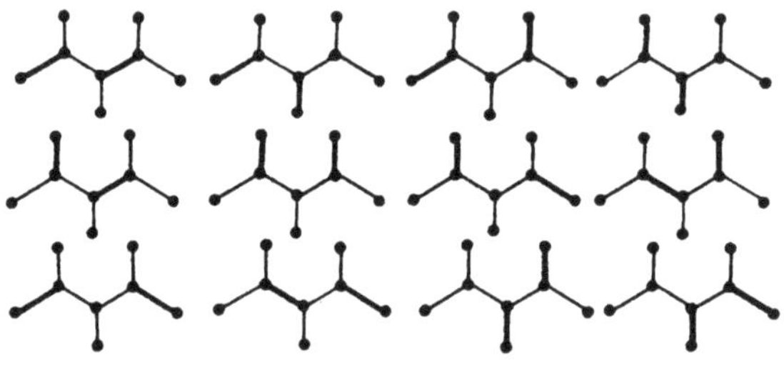

$p(G_1, 3) = 4$

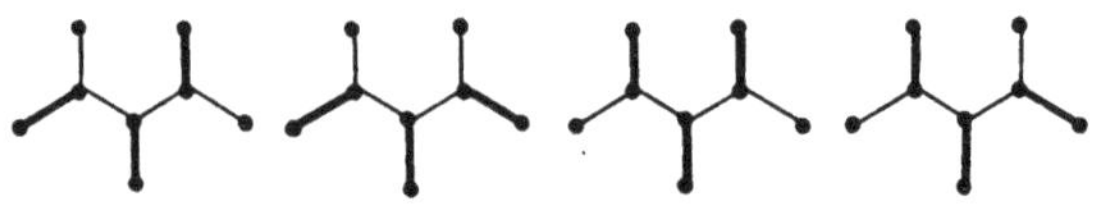

$p(G_1, 4) = 0$
Hence, $Z = 1 + 7 + 12 + 4 = 24$

For $\mathbf{G}_2$ one obtains: $p(G_2, 1) = 10$, $p(G_2, 2) = 33$, $p(G_2, 3) = 42$, $p(G_2, 4) = 18$, $p(G_2, 5) = 2$, and $Z = 105$.

Alternatively, for acyclic graphs Z can be defined as the sum of the absolute values of coefficients in the characteristic polynomial $P_H(G, x)$:

$$P_H(G, x) = \sum_{k=0}^{s} (-1)^k p(G, k) x^{N-2k} = (-1)^N \det(A - xE) \quad (39)$$

N is the number of atoms, A and E are the adjacency and unit matrix, respectively, while s is the largest number of edges disconnected to each other in the acyclic graph. Thus, for G_1 one obtains:

$$P_H(G_1, x) = \sum_{k=0}^{3} (-1)^k p(G_1, k) x^{8-2k} = x^8 - 7x^6 + 12x^4 - 4x^2$$

The characteristic polynomial may be obtained readily from the adjacency matrix A by placing x's on the main diagonal and expanding the determinant (39).

We classified Z in the group of indices associated with the distance matrix (and not to the adjacency matrix, as Trinajstić did [21,22]) due to the procedure for counting p(G, k).

The Hosoya index was applied [2,60-63] to correlations with boiling points, entropies, calculated bond orders, as well as for coding of chemical structures.

2.10 The Distance Polynomial

Analogously to the characteristic polynomial derived from adjacency matrix A, Hosoya et al. [64] introduced the distance polynomial:

$$P_D(G, x) = (-1)^N \det(D - xE) = \sum_{n=0}^{N} a_n(G) x^{N-n} \quad (40)$$

derived from the distance matrix D. The distance polynomial index:

$$Z' = \sum_{n=0}^{N} |a_n| \quad (41)$$

can be introduced similarly to the Hosoya index Z. It is conjectured that both $P_D(G, x)$ and Z' could characterize uniquely the graph, due to the very large a_n-values.

3 Centric Topological Indices

3.1 The Balaban Centric Indices

A new class of topological indices was devised for acyclic graphs (tree) by Balaban [65] on the basis of the notion of the graph centre. A vertex i is a centre of a graph if its maximal distance d_{ij} to any of the other vertex j in the graph is minimal. It is known from graph theory (the Jordan theorem) that any acyclic graph (tree) has a unique centre (vertex) or bicentre (two adjacent vertices). On pruning (lopping) a tree-graph towards its centre one obtains a sequence of numbers δ_i ("pruning sequence") of the vertices deleted at each step. A centric topological index B was defined by means of a quadratic formula, similar to that of the Zagreb index M_1:

$$B = \sum_{i=1}^{N} \delta_i^2 \tag{42}$$

A normalized centric index C was also introduced in order to reflect the topological shape of alkanes. By definition, C = 0 for n-alkanes (chain-graph):

$$C = \frac{1}{2}(B_{tree} - B_{chain}) = \frac{1}{2}(B_{tree} - 2N + U) \tag{43}$$

where $U = [1 - (-1)^N]$.

A further normalization (binormalization) was also made so as to have an upper limit of the centric index C equal to one for the case of star-graphs (i.e., graphs consisting of one vertex of degree N-1 and N-1 verties of degree N):

$$C' = \frac{C_{tree}}{C_{star}} = \frac{B - 2N + U}{(N-2)^2 - 2 + U} \tag{44}$$

The centric index C belongs to the "quadratic" class of indices, together with M_1 and N_2 indices. If the number of vertices of degree v_i is denoted by V_i, one obtains for graphs with $v_i \leqq 4$:

$$M_1 = 16V_4 + 9V_3 + 4V_2 + V_1 = 2(3V_4 + V_3) + 4N + 6 \tag{44'}$$

$$N_2 = 6V_4 + 3V_3 + V_2 = 3V_4 + V_3 + N - 2 \tag{44''}$$

Balaban normalized [65] similarly, for alkanes, the M_1 index specifying the normalized quadratic index Q:

$$Q = 4V_4 + \frac{3}{2}V_3 + 1 - \frac{1}{2}V_1 = 3V_4 + V_3 = \frac{1}{2}\sum_i iV_i - 2N + 3 \tag{45}$$

as well as the binormalized quadratic index Q':

$$Q' = \frac{3V_4 + V_3}{2(N-2)(N-3)} \tag{46}$$

The interrelations between Q, M_1 and N_2 are:

$$Q = M_1/2 - 2N + 3 = N_2 - N + 2 \tag{47}$$

$$M_1 = 2(N_2 + N - 1) \tag{48}$$

The pruning procedure, as well as the values of the five topological indices introduced by Balaban are illustrated below for $\mathbf{G_1}$:

For the standard linear C_8-alkane ($\mathbf{G_0}$) one obtains:

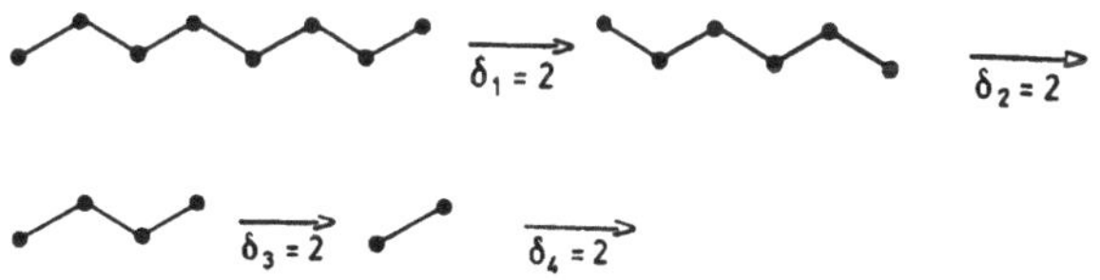

The values of the five quadratic indices are as:

$$B = 5^2 + 2^2 + 1^2 = 30; \quad C = (30 - 16)/2 = 7;$$
$$Q = 0 + 3 = 3; \quad C' = (30 - 2 \times 8 + 0)/[(8 - 2)^2 - 2 + 0] = 0.412$$
$$Q' = 3/[2\,(8 - 2)\,(8 - 3)] = 0.05$$

For $\mathbf{G_0}$ one obtains the value B = 16.

The centric indices correlated the best out of the known topological indices with the octane number of alkanes [66].

3.2 The Generalized Graph Centre Concept

The classical definition for the graph centre is not helpful for cyclic graphs where often a large number of vertices (most of them topologically non-equivalent) appear as central. Recently, Bonchev, Balaban and Mekenyan [67] proposed a generalized concept for the graph centre, and several centric indices were derived on this basis. The new definition consists of four hierarchically ordered criteria based on the distance matrix: 1) the smallest maximum distance in the row or column of the vertex; 2) the smallest sum of all distances d_{ij} in the row or column corresponding to vertex i; 3) the lowest number of largest entries in the vertex distance code, VDC; 4) criteria 1)—3) are repeated to the pseudocentre of the graph (obtained by removing the vertices discarded by the first three criteria and their incident edges) until on two successive cycles one and the same subgraph (i.e., the polycentre of the graph) is obtained. Further generalization of the graph centre concept was made by using additional criteria based on paths [68] and alternatively, by using an iterative procedure of mutual redetermination of the centric distance partitions of the graph vertices and edges [69]. The new concept of the graph centre could find many applications to the coding and rational nomenclature of chemical structures [70], the classification and coding of chemical reaction mechanisms [71], correlations with proton chemical shifts [68], etc.

4 Indices Based on Information Theory

Information theory [71,72] is a convenient basis for the quantitative characterization of structures. It introduces simple structural indices called information content (total or mean) of any structured system. For such a system having N elements distributed into k classes of equivalence $N_1, N_2, \dots, N_k$ a probability distribution $P\{p_1, p_2, \dots, p_k\}$ is constructed ($p_i = N_i/N$). The entropy of this distribution, calculated [71] by the Shannon formula[1]:

$$\bar{I} = \sum_i p_i \text{ lb } p_i \tag{49}$$

is called mean information content of the structure, while a derived expression [72] specifies the total information:

$$I = N\bar{I} = N \text{ lb } N - \sum_i N_i \text{ lb } N_i \tag{50}$$

The equations (49) and (50) can also be used for the case when N is a non-integer quantity partitioned into k different contributions N_i.

$\bar{I}$ and I are generally called information indices, name additionally specified according to the equivalence relation used to generate the probability distribution (for detailed reviews one may consult refs. [74,75]).

4.1 Information on the Graph Orbits

Information indices can be defined on any structural basis, topological or non-topological one.

The first topological-information index was introduced by Rashevsky [76] and Trucco [77] in 1955–1956, on the basis of the automorphism group of the graph. Two vertices belong to the same graph orbit if there is an automorphism (i.e., a permutation preserving adjacency) that maps one onto the other. Denoting the cardinality of the i-th orbit by N_i, and constructing thus the orbital partition of the

[1] $\text{lb } x = \log_2 x$

graph vertices, one arrives by means of Eqs. (49) and (50) to the topological information (as Rashevsky called it), or more precisely, the information on the graph orbits, $\bar{I}_{ORB}$. For the graph G_1 one obtains:

orbit I: vertices {1, 5, 6, 8}, $N_1 = 4$
orbit II: vertices {7}, $N_2 = 1$
orbit III: vertices {2, 4}, $N_3 = 2$
orbit IV: vertices {3}, $N_4 = 1$

Using the symbol P for partition, we have:

$$P_{ORB} = N\{N_1, N_2, N_3, N_4\} = 8\{4, 1, 2, 1\}, \text{ and } \bar{I}_{ORB} = 1.7500 \text{ bits.}$$

For the graph G_2 we have:

orbit	1	2	3	4	5	6	7	8
vertices	1	2,6	3,5	4	7	8	9	10
cardinality	1	2	2	1	1	1	1	1

$$P_{ORB} = 10\{1, 2, 2, 1, 1, 1, 1, 1\}, \text{ and } \bar{I}_{ORB} = 2.9219 \text{ bits.}$$

The Rashevsky information index was applied to the search for an information balance of chemical reactions [78], and to the study of the problem of selfgeneration of life on Earth [79]. The properties of this index were studied by Mowshowitz [80], who also introduced a second topological information index based on the colouring of the graph.

4.2 The Chromatic Information Index

A colouring of a graph is an assignment of k colours to the graph vertices in such a way that no two adjacent vertices have the same colour. Hence, a chromatic partition of the graph vertices into colour classes can be obtained. This partition in general is not unique, but Mowshowitz defined [80] a unique chromatic information content of the graph, $\bar{I}_{CHR}$, which reflects the chromatic structure of the graph. This was done by means of minimization of the Shannon expression over a certain class of chromatic partitions V:

$$\bar{I}_{CHR} = \min V \left\{ -\sum_{i=1}^{k} \frac{N_i}{N} \operatorname{lb} \frac{N_i}{N} \right\} \tag{51}$$

The colouring of the vertices of bicolourable graphs G_1 and G_2 is unique:

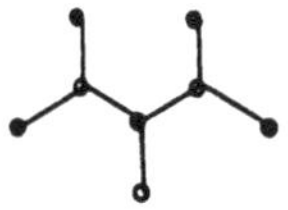

$N_1 = 5,$ $N_2 = 3,$ $P_{CHR} = 8\{5, 3\},$ and $\bar{I}_{CHR} = 0.9544$ bits

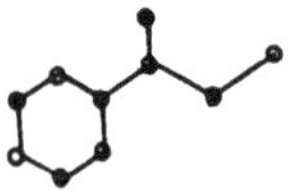

$N_1 = 5,$ $N_2 = 5,$ $P_{CHR} = 10\{5, 5\},$ and $\bar{I}_{CHR} = 1.0000$ bits

4.3 Information Indices on the Graph Distances

Two types of information indices resulted from the statistical analysis of the distance matrix D(G) made by Bonchev and Trinajstić [34]. Proceeding from Eq. (26), one can come to two distance partitions. In the first one, P_D^E, the total number of distances is partitioned into classes of distances, according to their equality or non-equality:

$$P_D^E = \frac{N(N-1)}{2}\{g_1, g_2, \dots, g_{d\,max}\} \tag{52}$$

while in the second one, P_D^W, the total distance (or, the Wiener index) is partitioned into different individual distances d_i:

$$P_D^W = w\{g_1d_1, g_2d_2, \dots, g_md_{max}\} \tag{53}$$

The respective information indices for the equality of distances, $\bar{I}_D^E$, and for the magnitude of distances, $\bar{I}_D^W$, are calculated by the equations:

$$\bar{I}_D^E = -\sum_i \frac{2g_i}{N(N-1)} \operatorname{lb} \frac{2g_i}{N(N-1)}, \quad \text{bits} \tag{54}$$

$$\bar{I}_D^W = -\sum_i g_i \frac{i}{W} \operatorname{lb} \frac{i}{W}, \quad \text{bits} \tag{55}$$

The calculations for graphs G_1 and G_2 give:

$P_D^E = 28\{7. 9, 8, 4\}$, $\bar{I}_D^E = 1.9438$ bits
$P_D^W = 65\{6\times1, 9\times2, 8\times3, 4\times4\}$, $\bar{I}_D^W = 4.6679$ bits

and, respectively:

$P_D^E = 45\{10, 12, 11, 7, 4, 1\}$, $\bar{I}_D^E = 2.3375$ bits
$P_D^W = 121\{10\times1, 12\times2, 11\times3, 7\times4, 4\times5, 1\times6\}$, $\bar{I}_D^W = 5.3135$ bits

The information indices have, in general, a greater power for discriminating isomers than the respective topological indices, due to the non-integer values of the information function. This is particularly true for the above two indices as compared with the Wiener index. Because of this, the two information indices on the graph distances are a powerful tool for characterizing molecular branching and cyclicity [34], and correlate very well with thermodynamic properties [81] and the gas-chromatographic retention time [82].

4.4 Information Analogous with the Randić and Hosoya Indices

The distribution of the Hosoya Z index into its constituent non-adjacent numbers p(G, k):

$$P_Z = Z\{p(G, 0), p(G, 1), \ldots, p(G, [N/2])\} \qquad (56)$$

was used by Bonchev and Trinajstić [34] to define the information index $\bar{I}_Z$ on the Hosoya graph decompositions:

$$\bar{I}_Z = -\sum_{k=0}^{[N/2]} \frac{p(G, k)}{Z} \operatorname{lb} \frac{p(G, k)}{Z}, \quad \text{bits} \qquad (57)$$

The calculations for graphs G_1 and G_2 give:

$P_Z = 24\{1, 7, 12, 4\}$, $\bar{I}_Z = 1.6403$ bits, respectively
$P_Z = 106\{1, 10, 33, 42, 18, 2\}$, $\bar{I}_Z = 1.9806$ bits.

The information analogue of the Randić index, χ_R, was defined [83] by partitioning the graph edges into k classes, depending on the equality of their partial connectivity indices χ_i (i.e., $\chi_R = \sum_i \chi_i$). This information index on the distribution of edges in the graph according to their equivalence, $\bar{I}_\chi^E$, is given by:

$$\bar{I}_\chi^E = -\sum_{i=1}^{k} \frac{M_i}{M} \operatorname{lb} \frac{M_i}{M}, \quad \text{bits} \qquad (58)$$

where M_i is the number of edges having the same partial connectivity index (note that $M = \sum_i^k M_i$).

For the graphs G_1 and G_2 we have:

$P_\chi^E = 7(5, 2)$, $\bar{I}_\chi^E = 0.8631$ bits, respectively
$P_\chi^E = 10(1, 1, 1, 3, 4)$, $\bar{I}_\chi^E = 2.0464$ bits

4.5 The Information Centric Indices

The topological centric indices [65,67] can be expressed both in quadratic and logarithmic (information) form. In general, each partition of the quantity X into a set of constituents X_i can be converted into quadratic, Q, or information index, $\bar{I}$, as:

$$P_X = X\{X_1, X_2, \dots, X_k\} \tag{59}$$

$$Q = \sum_{i=1}^{k} X_i^2 \tag{60}$$

$$\bar{I} = \sum_{i=1}^{k} \frac{X_i}{X} \operatorname{lb} \frac{X_i}{X} \tag{61}$$

4.6 The Orbital Information Index for the Graph Connections

Bertz [83a)] used the number of connections (defined as the number of pairs of adjacent edges) to introduce the index C(η), called the complexity of a molecule.

In simple graphs, i.e., graph representing molecules without multiple bonds, the connections are paths of length two, or propane fragments used for the derivation of N_2 index. For these cases the Bertz index is an information analogue, $\bar{I}_{N_2}$, to the index N_2. Re-entrant paths of length two along double bonds take into account the presence of multiple bonds. Alternatively, the connections may be defined as the number of lines in the bond graph [83b)]: if each edge of the original molecular graph is represented by a point of the bond graph and if two such points are connected wherever the corresponding edges are adjacent, the bond graph is obtained.

In order to use the orbital distribution of the graph connections as measure for molecular complexity, Bertz modified the Shannon equation (i.e., the Mowshowitz [80)] variant) adding the term N lb N:

$$C(\eta) = 2N \operatorname{lb} N - \sum_i N_i \operatorname{lb} N_i \tag{62}$$

In order to take into account the increased complexity of molecules with heteroatoms (i.e., graphs with coloured vertices), Bertz combined C(η) with the information for chemical composition, I_{CC}, introduced by Bra..son [84a)] and by Dancoff and Quastler [84b)], resulting in:

$$C_T = C(\eta) + I_{CC} \tag{63}$$

As an example we shall take the graph $\mathbf{G}_1$ modified by the presence of two double and two heteroatoms denoted by black vertices:

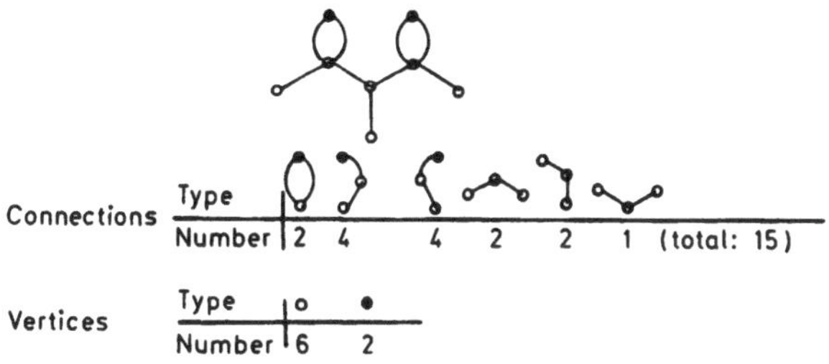

$$C(\eta) = 2 \times 15 \text{ lb } 15 - 2 \times 4 \text{ lb } 4 - 3 \times 2 \text{ lb } 2 - 1 \ln 1 = 95.21 \text{ bits},$$
$$I_{CC} = 8 \text{ lb } 8 - 6 \text{ lb } 6 - 3 \text{ lb } 2 = 19.42 \text{ bits, and}$$
$$C_T = 95.21 + 10.42 = 114.63 \text{ bits}.$$

The $C(\eta)$ index was applied [83] to quantitative estimates of the change in molecular complexity during a chemical reaction in relation to computer-assisted molecular design. A similar approach was reported in 1980 by Dosmorov [85] who used a combined information index as a maximizing criterion for predicting the preferable reaction path in complex organic syntheses. This author included in his index $\bar{I}_D^W$, I_{at} (the atomic information content of the chemical elements [86c]), I_{sym} (the index on molecular symmetry [86a]), and I_{CC} (the index of chemical composition [86b]).

One should mention here that I_{at} is a missing term in the complexity C_T which will result in the same C_T-values for molecules having chemically different substituents of the same valency (e.g., F, Cl, Br, I). Taking into account the results of Sarkar, Roy and Sarkar [87], and Rashevsky [76] we conclude that the Bertz approach is interesting in both theoretical and practical aspects, but the title "The first general index of molecular complexity" is somewhat exaggerated.

4.7 The Topological Information Superindex

A comparative analysis of the discrimination power of eleven indices (see Table 1, all indices but J) indicated [88] that no index so far described discriminates isomers uniquely.

A combined topological index, named superindex, SI, consisting of a number of topological indices was proposed.[88] Three versions of SI have been presented: i) the sum of the first ten indices in Table 1; ii) SI $= \{\bar{I}_D^W, \bar{I}_X^E, \bar{I}_Z, \bar{I}_C, \bar{I}_{ORB}, \bar{I}_{CHR}\}$; iii) the entries of the SI are expressed in terms of sets of integers which are the partitions used in the determination of these indices.

Considering the very low degeneracy of index J (the last one in Table 1) in comparison with the other topological indices, a fourth approach seems possible, namely the summation of only two topological indices, J and χ, which provides complete discrimination of all known structures.

Table 1. Mean degeneracy of eleven topological or topological information indices calculated for 427 acyclic, monocyclic and bicyclic simple graphs with 4 to 8 vertices

No. of vertices in the isomer set	No. of graphs	w	I_D^E	I_D^W	Z	I_Z	χ_R	I_χ	I_C	I_{CHR}	I_{ORB}	J
(Acyclic graphs)												
4	2	1	1	1	1	1	1	1	1	1	1	1
5	3	1	1	1	1	1	1	1	1	1.5	1	1
6	5	1	1	1	1	1	1	1	1	2	1.2	1
7	11	1.22	1	1	1	1	1	1	1.8	3.7	1.2	1
8	24	1.15	1	1	1.28	1	1.09	1	2.3	5.8	1.5	1
(Mono and bicyclic graphs)												
4	3	1.5	1.5	1.5	1	1	1	1	1	1.5	1	1
5	10	2.5	2	2	1.7	1.4	1	1.4	2	3.3	2	1
6	29	3.6	2.4	2.2	2.6	1.5	1.2	1.6	4.1	5.8	3.6	1
7	85	6	2.8	2.4	4.5	2.2	1.6	3.4	12.1	12.1	9.4	1
8	255	10.6	3.5	2.8	7.7	2.2	1.6	6	16	36.4	18.2	1.008

4.8 The Electropy Index

I'Haya et al. [89] introduced an information index ε, called electropy, based on the assumption that the molecule forms a finite space Γ which is divided into several partial bond spaces according to the electronic pairings in the molecule. The electropy, ε, is viewed as a measure of the degree of freedom of choice for electrons in occupying different partial spaces in Γ during the process of molecular formation.

The calculation of electropy is carried out according to the equation of information theory [73] used in cases of equal probabilities of the possible events P_0 and P_1:

$$\varepsilon = \mathrm{lb}\,(P_0/P_1) \tag{64}$$

where the number of a priori and a posteriori possible events, corresponding to those before and after formation of a molecule, respectively, is denoted by P_0 and P_1; $P_1 = 1$, since the distribution of electrons is unique for each definite molecular structure. Taking into account that P_1 is the total number of possible ways of distributing N particles into k partial bond spaces with N_i particles in the partial space i, the final formula is obtained:

$$\varepsilon = \mathrm{lb}\left(N!\Big/\prod_{i=1}^{k} N_i!\right) \tag{65}$$

One can calculate the electropy of graph G_1 as follows:

$\Gamma_{1s(C)}$	for 8 carbon atoms:	$2\times 8 = 16$
Γ_{C-C}	for 7 C—C bonds:	$14\times 1 = 14$
Γ_{C-H}	for 3 CH groups:	$2\times 3 = 6$
Γ_{CH_2}	for 0 CH_2 groups:	0
Γ_{CH_3}	for 5 CH_3 groups:	$6\times 5 = 30$
		(total: 66)

Therefore, $\varepsilon(G_1) = \mathrm{lb}\,[66!/(2!)^8\,(14!)^1\,(2!)^3\,(6!)^5] = 213.3$ bits.

A correction term to electropy is introduced which takes into account the effect of steric hindrance between adjacent methyl groups connected to the t-th carbon atom:

$$\mathrm{lb}\,\zeta = \sum_{t=1}^{N} \mathrm{lb}\,\zeta_t$$

where N stands for the number of carbon atoms, $\zeta_t = 1$ when there is no steric hindrance, and $\zeta_t = 2$, 6 or 12 when two, three or four methyl groups are connected to one carbon atom.

For instance, in the case of the above graph G_1, $\mathrm{lb}\,\zeta = \mathrm{lb}\,2 + \mathrm{lb}\,2 = 2$ bits, therefore $\varepsilon'(G_1) = \varepsilon(G) - \mathrm{lb}\,\zeta = 211.3$ bits.

The combination of the electronic and topological factors within this approach yields high correlations with thermodynamic properties of molecules.

Through the I'Haya electropy was meant as an electronic information index, it is essentially topological in nature, but takes also into account electronic and steric factors through the intermediacy of topology.

4.9 The Merrifield and Simmons Indices

Recently, Marrifield and Simmons [92] argued that the total number of the following subsets of the vertices or edges of a graph **G**:
i) the independent set of vertices, $\sigma(\mathbf{G})$, or edges, $\bar{\sigma}(\mathbf{G})$;
ii) the connected set of vertices, $\varrho(\mathbf{G})$, or edges, $\bar{\varrho}(\mathbf{G})$;
iii) the point cover, $\chi(\mathbf{G})$, or line cover, $\bar{\chi}(\mathbf{G})$;
iv) the externally stable vertex, $\varepsilon(\mathbf{G})$, or edge set, $\bar{\varepsilon}(\mathbf{G})$; and
v) the irredundant vertex, $\iota(\mathbf{G})$, or edge set, $\bar{\iota}(\mathbf{G})$,
are structure-sensitive graphical subsets. Because each of these subsets is an expression of the adjacency or incidence relations within the graph **G**, the total number of these subsets (i.e., σ, $\bar{\sigma}$, ϱ, $\bar{\varrho}$, χ, $\bar{\chi}$, ε, $\bar{\varepsilon}$, ι and $\bar{\iota}$) may be viewed as topological indices. They are quite varied in the manner in which they quantify various aspects of the graph structure.

The inspection of the results obtained for all connected graphs of up to six vertices motivates the conclusions collected in Table 2.

These authors showed [92] that σ and $\bar{\sigma}$ are sensitive to chemically significant details of the molecular structure and that the bond strength pattern within a molecule correlates with the numbers of independent sets intersecting the different bonds.

Table 2. Qualitative properties of subset counts

	Branching	Cyclization
σ, χ	Increases	Decreases
ϱ	Increases	Increases
ε	Variable	Increases
ι	Variable	Variable
$\bar{\sigma}$	Decreases	Increases
$\bar{\varrho}$	Increases	Increases
$\bar{\chi}$	Decreases	Increases
$\bar{\varepsilon}$	Increases	Increases
$\bar{\iota}$	Decreases	Increases

5 Concluding Remarks

A review on the topological indices can hardly be complete. We had to omit here some indices like structure count (SC) and algebraic structure count of Herndon [91] which are related with the characterization of π-electron molecules.

The large number of existing topological indices raises the question to what extent they are orthogonal. In other words, it is possible that some existing topological indices express predominantly the same type of structural information; the differences then reside in the scaling factors. We computed the linear relationships between pairs of topological indices T_i and T_j:

$$T_i = a + bT_j$$

For avoiding chance correlations, three series of hydrocarbons with wide structural variations were considered: 37 alkanes, 36 polyalkylcyclohexanes and 48 monocyclic structures (the full results are reported elsewhere [93a)]).

The strongest intercorrelations among investigated topological indices are shown in Fig. 2A.

From their definitions, one may admit that topological indices may code two structural factors, namely, the size and the shape of the molecule. This assumption is proved by calculating [94)] the schemes (B) in Figure 2 for the 36 polyalkylcyclohexanes with 6–10 carbon atoms, and for a subseries of 22 polyalkylcyclohexanes with 10 carbon atoms (resulting in scheme (C)). The schemes (A)–(C) corroborated with the

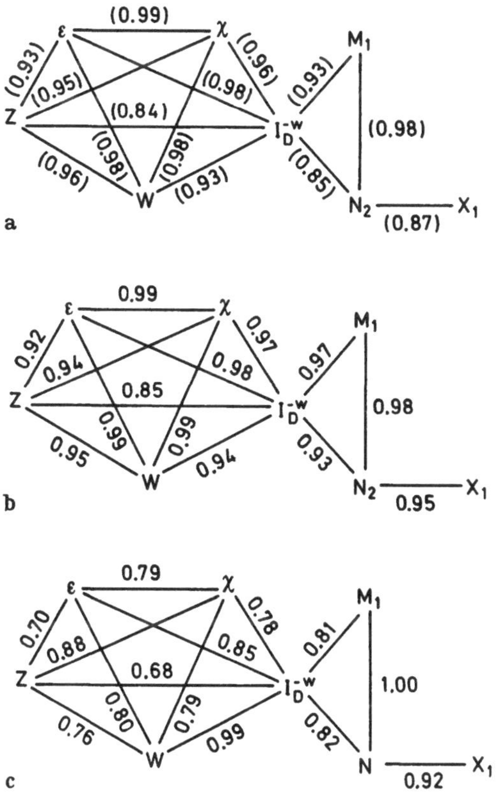

Fig. 2a–c. The strongest intercorrelations among investigated topological indices. **a** average value for correlation coefficients obtained in the three series of structures; **b** correlation coefficients for the polyalkylcyclohexanes with 6–10 carbon atoms; **c** correlation coefficients for the polyalkylcyclohexanes with 10 carbon atoms

correlations T_i vs. number of carbon atoms in a series of 37 alkanes [93a] (the number of carbon atoms represents an approximate measure of the molecular size):

T_i	M_1	$\bar{I}_D^w$	χ	ε	w	Z
correlation coefficient	0.92	0.99	0.97	0.99	0.96	0.89

argue that the dominant component within topological indices M_1, $\bar{I}_D^w$, χ, ε, w and Z is the volume component. Accordingly, these indices should be viewed as bulk steric parameters. The meaning of the other topological indices is rather obscure.

As possible future trends in the development of the area of topological indices and of their uses, one may cite:

i) The deeper exploration of the idea that "normalization" can result in quantifying the shapes of molecules. Indeed, most of the topological indices so far described consist of a shape and a bulk component in unknown proportions. If one normalizes a topological index, excluding thereby the bulk component, one may combine the remaining "shape" or "branching" component with the "size" or "bulk" component (expressed adequately, e.g. by the number of vertices, by molar refraction [93b], etc.) in the proportion required by the phenomenon under consideration.

ii) Till now, a handicap in using topological indices for QSAR has been the lack of adequate provisions for specifying the nature of graph vertices (i.e., the identity of the atoms) or of graph edges (i.e., the chemical bonds). Similarly, for biologically active molecular fragments, the topological index approach usually fails to discriminate the site of attachment (treating, e.g. s-butyl and n-butyl groups in the same manner and yielding therefore the same topological index). It is to be hoped that an appropriate treatment of these aspects will wrap the dry mathematical bones of most present-day topological indices in chemical garments adequate for many applications.

iii) Until now, research in the field of topological indices consisted in adopting concepts from mathematics and topology which could benefit chemistry. It is conceivable that this trend could be partly reversed in the future, since with the aid of highly discriminating indices one may tackle old unsolved graph-theoretical problems such as recognition of graph isomorphism, partitioning of graph vertices into automorphic orbits, finding all isomers with a given formula (the last problem is of equal interest for chemistry and graph theory).

iv) Since index J is singled out by its very low degeneracy on the one hand, and by the fact that it quantifies practically only the shape of the molecule, it provides interesting possibilities for steric mapping of biological receptors. A co-operative project is under way for improving the receptor mapping by means of SIBIS-type algorithms [95], which use J as a variable to code the molecular branching, which should not be ignored.

6 Acknowledgements

Thanks are addressed to Professors O. E. Polansky, N. Trinajstić and S. V. Dosmorov for making available unpublished information.

7 References

1. Bonchev, D., Mekenyan, O., Fritsche, J.: Cryst. Growth *49*, 90 (1980)
2. Burton, J. J.: In: Sintering and Catalysis, (ed. Kuczynski, G. C.) Plenum, New York 1977, p. 17–27
3. Gutman, I. et al.: J. Chem. Phys. *62*, 3339 (1975)
4. Gutman, I., Trinajstić, N.: Chem. Phys. Lett. *17*, 535 (1972)
5. Randić, M.: J. Amer. Chem. Soc. *97*, 6609 (1975); Int. J. Quantum Chem., Symp. *5*, 245 (1978)
6. Kier, L. B. et al.: J. Pharm. Sci. *64*, 1971 (1975)
7. Kier, L. B. et al.: J. Pharm. Sci. *65*, 1226 (1976)
8. Kier, L. B., Hall, L. H.: Molecular Connectivity in Chemistry and Drug Research, Academic, New York 1976
9. DiPaolo, T., Kier, L. B., Hall, L. H.: Mol. Pharmacol. *13*, 31 (1977)
10. Nizshnii, S. V., Epstein, N. A.: Usp. Khim. *47*, 739 (1978)
11. DiPaolo, T.: J. Pharm. Sci. *67*, 564, 566 (1978)
12. DiPaolo, T., Kier, L. B., Hall, L. H.: J. Pharm. Sci. *68*, 39 (1979)
13. Balaban, A. T. et al.: Steric Fit in QSAR, Lecture Notes in Chemistry vol. 15, Springer, Berlin 1980
14. Murray, W. T., Hall, L. H., Kier, L. B.: J. Pharm. Sci. *64*, 1978 (1975)
15. Kier, L. B. et al.: J. Med. Chem. *18*, 1272 (1975)
16. Richard, A. J.,Kier, L. B.: J. Pharm. Sci. *69*, 124 (1980)
17. Riet, B. van't, Kier, L. B., Elford, H. L.: J. Pharm. Sci. *69*, 586 (1980)
18. Platt, J. R.: J. Chem. Phys. *15*, 419 (1947); J. Phys. Chem. *56*, 328 (1952)
19. Gordon, M., Scantlebury, G. R.: Trans. Faraday Soc. *60*, 605 (1964)
20. Bonchev, D., Knop, J. V., Trinajstić, N.: Math. Chem. *6*, 21 (1979)
21. Sabljć, A., Trinajstić, N.: J. Pharm. Eng. (in press)
22. Trinajstić, N.: Chemical Graph Theory, New York, Chemical Rubber Co., Boca Raton, 1983
23. Gutman, I., Randić, M.: Chem. Phys. Lett. *47*, 15 (1977)
24. Muirhead, R. F.: Proc. Edinburgh Math. Soc. *19*, 36 (1901); *21*, 144 (1903); *24*, 45 (1906)
25. Randić, M.: J. Chem. Phys. *60*, 3920 (1974); *62*, 309 (1975); J. Chem. Inf. Comp. Sci. *15*, 105 (1975); Chem. Phys. Lett. *42*, 283 (1976)
26. Lovasz, L., Pelikan, J.: Period. Math. Hung. *3*, 175 (1973)
27. Cvetković, D., Gutman, I.: Croat. Chem. Acta *49*, 115 (1975)
28. Wiener, H.: J. Amer. Chem. Soc. *69*, 17, 2636 (1947); J. Chem. Phys. *15*, 766 (1947); J. Chem. Phys. *52*, 425, 1082 (1948)
29. Hosoya, H.: Bull. Chem. Soc. Japan *44*, 2332 (1971)
30. Rouvray, D. H.: Math. Chem. *1*, 125 (1975)
31. Rouvray, D. H., Crafford, B. C.: South African J. Sci. *72*, 47 (1976)
32. Bonchev, D. et al.: J. Chromatogr. *176*, 149 (1979)
33. Trinajstić, N. et al.: Kem. Ind. (Zagreb) *28*, 527 (1979)
34. Bonchev, D., Trinajstić, N.: J. Chem. Phys. *67*, 4517 (1977)
35. Bonchev, D., Trinajstić, N.: Int. J. Quantum Chem., Symp., *12*, 293 (1978)
36. Bonchev, D.: Croat. Chem. Acta *52*, 361 (1979)
37. Mekenyan, O., Bonchev, D., Trinajstić, N.: Math. Chem. *6*, 93 (1979)
38. Bonchev, D., Mekenyan, O., Trinajistić, N.: Int. J. Quantum Chem. *17*, 845 (1980)
39. Mekenyan, O., Bonchev, D., Trinajistić, N.: Int. J. Quantum Chem. *19*, 929 (1981); Math. Chem. *11*, 145 (1981)
40. Bonchev, D., Mekenyan, O., Fritsche, H.: J. Cryst. Growth *49*, 90 (1980)
41. Bonchev, D., Mekenyan, O., Fritsche, H.: Phys. Stat. Sol. A *55*, 181 (1979)
42. Mekenyan, O., Bonchev, D., Fritsche, H.: Phys. Stat. Sol. A *56*, 607 (1979)
43. Bonchev, D., Mekenyan, O.: Z. Naturforsch. *35A*, 739 (1980)
44. Bonchev, D., Mekenyan, O., Polansky, O. E.: Z. Naturforsch. *36A*, 643, 647 (1981)
45. Mekenyan, O., Dimitrov, C., Bonchev, D.: J. Polym. Sci. (submitted)
46. Polansky, O. E.: unpublished
47. Bonchev, D., Balaban, A. T., Mekenyan, O.: J. Chem. Inf. Comp. Sci. *20*, 106 (1980)

47a. Balaban, A. T.: Chem. Phys. Lett. *89*, 399 (1982)

48. Bernstein, H. J.: J. Chem. Phys. *19*, 140 (1951); ibid. *20*, 263, 351 (1952); J. Phys. Chem. *69*, 1550 (1965)
49. Tatevskii, V. M.: Khimicheskoe stroenye uglevodorodov: zakonomernosti v ikh fiziko-khimicheskikh svoisto, Moscow 1953; Tatvskii, V. M., Benderskii, V. A., Yarovoi, S. S.: Metody rascheta fiziko-khimicheskikh svoistv parafinovykh uglevodorodov, Moscow 1953
50. Tatevskii, V. M., Papulov, Yu. G.: Zhur. Obsh. Khim. *34*, 241, 489, 708 (1960); Papulov, Yu. G. et al.: Zhur. Fiz. Khim. *48*, 31 (1974)
51. Smolenskii, E. A.: Zhur. Fiz. Khim. *38*, 1288 (1964)
52. Altenburg, K.: a) Kolloid Z. *178*, 112 (1961);
b) Brennstoff Chem. *47*, 100, 331 (1966);
c) Z. Phys. Chem. (Leipzig) *261*, 389 (1980)
53. Randić, M.: Math. Chem. *7*, 5 (1979)
54. Randić, M., Wilkins, C. L.: J. Chem. Inf. Comp. Sci. *19*, 31 (1979)
55. Randić, M., Wilkins, C. L.: Chem. Phys. Lett. *63*, 332 (1979); J. Phys. Chem. *83*, 1525 (1979)
56. Randić, M.: J. Chem. Inf. Comp. Sci. *18*, 101 (1978)
57. Randić, M., Wilkins, C. L.: J. Chem. Inf. Comp. Sci. *20*, 36 (1980)
58. Randić, M.: J. Magn. Reson. *39*, 431 (1980)
59. Bonchev, D.: unpublished
60. Hosoya, H.: J. Chem. Doc. *12*, 181 (1972); Fibonacci Quart. *3*, 255 (1973)
61. Hosoya, H., Kawasaki, K., Muzutani, K.: Bull. Chem. Soc. Japan *45*, 3415 (1972); Narumi, H., Hosoya, H.: ibid *53*, 1228 (1980)
62. Mizutani, K., Kawasaki, K., Hosoya, H.: Nat. Sci. Rept. Ochanomizu Univ. *22*, 39 (1971)
63. Kawasaki, K., Mizutami, K., Hosoya, H.: Nat. Sci. Rept. Ochanomizu Univ. *22*, 181 (1971)
64. Hosoya, H., Murakami, M., Gotoh, M.: Nat. Sci. Rept. Ochanomizu Univ. *24*, 27 (1973)
65. Balaban, A. T.: Theor. Chim. Acta *53*, 355 (1979)
66. Balaban, A. T., Motoc, I.: Math. Chem. *5*, 197 (1979)
67. Bonchev, D., Balaban, A. T., Mekenyan, O.: J. Chem. Inf. Comp. Sci. *20*, 106 (1980)
68. Bonchev, D., Balaban, A. T., Randić, M.: Int. J. Quantum Chem. *19*, 61 (1981)
69. Bonchev, D., Gruncarov, I.: to be published
70. Bonchev, D., Balaban, A. T.: J. Chem. Inf. Comp. Sci. (in press)
71. Bonchev, D., Tempkin, O. N., Kamenski, D.: J. Comp. Chem. (in press)
72. Shannon, C., Weaver, W.: Mathematical Theory of Communication, Urbana, Univ. Illinois Press, 1949
73. Brillouin, L.: Science and Information Theory, New York, Academic, 1956
74. Bonchev, D.: Math. Chem. *7*, 65 (1979)
75. Bonchev, D.: Theoretic Information Indices for Characterization of Chemical Structures, Chichester, Research Studies, 1983
76. Rashevsky, N.: Bull. Math. Biophys. *30*, 229 (1955)
77. Trucco, E.: Bull. Math. Biophys. *18*, 129, 237 (1956)
78. Karreman, G.: Bull. Math. Biophys. *17*, 279 (1955)
79. Rashevsky, N.: Bull. Math. Biophys. *22*, 351 (1960)
80. Mowshowitz, A.: Bull. Math. Biophys. *30*, 175, 225, 387, 533 (1968)
81. Mekenyan, O., Bonchev, D., Trinajstić, N.: Int. J. Quantum Chem., Symp. *18*, 369 (1980)
82. Papozova, D., Dimov, N., Bonchev, D.: J. Chromatogr. *188*, 297 (1980)
83. Bertz, S. H.: a) J. Amer. Chem. Soc. *103*, 3599 (1981);
b) J. Chem. Soc. Chem. Comm. 819 (1981)
84. a) Branson, H.: in: Essays on the Use of Information Theory in Biology, (ed. Quastler, H.) Univ. Illinois Press, Urbana 1953;
b) Dancoff, A., Quastler, H.: ibid.
85. Dosmorov, S. V.: Reports of the Fifth All-Union Conf. on the Use of Computers in Molecular Spectroscopy and Chemical Studies, p. 28, Novosibirsk, 1980 (in Russian); Kinet. Katal. (in press)
86. a) Bonchev, D., Kamenski, D., Kamenska, V.: Bull. Math. Biol. *38*, 119 (1976);
b) Bonchev, D., Peev, T.: Jahresber. Hochsch. Chem. Technol. Burgas *10*, 561 (1973);
c) Bonchev, D., Kamenska, V., Kamenski, D.: Monatsh. Chem. *108*, 477 (1977);
d) Bonchev, D.: Math. Chem. *7*, 65 (1979)
87. Sarkar, P., Roy, A. B., Sarkar, P. K.: Math. Biosci. *39*, 299 (1978)

88. Bonchev, D., Mekenyan, O., Trinajstić, N.: J. Comp. Chem. *2*, 127 (1981)
89. Yee, W. T., Sakamoto, K., I'Haya, Y. J.: Rept. Univ. Electro-comm. *27*, 53 (1976)
90. Sakamoto, K., Yee, W. T., I'Haya, Y. J.: Rept. Univ. Electro-Comm. *27*, 227 (1977)
91. Herndon, W. C.: Int. J. Quantum Chem. *QBS 1*, 123 (1974); Ann. N.Y. Acad. Sci. *36*, 200 (1974)
92. Merrifield, R. E., Simmons, H. E.: Proc. Natl. Acad. Sci. US *78*, 692, 1329 (1981); Theor. Chim. Acta *55*, 55 (1980)
93. a) Motoc, I. et al.: Math. Chem. *13*, 369 (1982)
b) Motoc, I., Balaban, A. T.: Rev. Roumaine Chim. *26*, 593 (1981)
94. Motoc, I.: unpublished
95. Motoc, I.: Arzneim. Forsch. *31*, 290 (1981); Motoc, I., Dragomir, O.: Math. Chem. *12*, 117 (1981)

The Upsilon Steric Parameter — Definition and Determination

Marvin Charton

Chemistry Department, School of Liberal Arts and Sciences, Pratt Institute, Brooklyn, NY 11205, USA

Table of Contents

The definition of steric parameters initially based on Van der Waals raddi (r_V) is described. Tables of various sets of r_V are given. These different types of r_V are shown to be interrelated. The r_V values of Bondi are chosen as the standard set. Substituents are classified according to the degree of conformational dependence of their steric effect. Values of the steric parameter υ are obtained directly from r_V whenever possible. For symmetric MZ_n groups they are obtained from calculated r_V values. For planar π-bonded groups equations are derived which permit the calculation of both υ and the corresponding delocalized electrical effect parameter. Values of υ for other groups are based on kinetic data scaled by correlation with groups for which υ is known. Over 300 values of υ are reported.

Abbreviations

Ac	acetyl	**Prefixes**	*Prefixes*
Ak	alkyl	i	iso
Am	amyl	s	secondary
Bu	butyl	t	tertiary
Bz	benzoyl	c	cyclo
Dd	dodecyl		
Et	ethyl	**Other**	*Other*
Hp	heptyl	X	substituent
Hx	hexyl	G	skeletal group
Hpd	heptadecyl	Y	active site
Me	methyl	NCD	no conformational dependence
No	nonyl	MCD	minimal conformational dependence
Oc	octyl	ICD	intermediate conformational dependence
Pnd	pentadecyl	LCD	large conformational dependence
Ph	phenyl	MSI	minimal steric interaction principle
Pr	propyl	r_V	van der Waals radius
Td	tridecyl	r_c	covalent radius
Ud	undecyl	υ_{ef}	effective steric parameter
Vi	vinyl	υ_{mx}	maximum steric parameter
		υ_{mn}	minimum steric parameter
		υ_{ax}	axial steric parameter

1 Introduction

1.1 The Method

A major method of modeling the effect of structural variation on chemical reactivity, physical properties or biological activity of a set of substrates is the use of correlation analysis. In this method it is assumed that the effect of structural variation of a substituent X upon some chemical, physical or biological property of interest is a linear function of parameters which constitute a measure of electrical, steric, and transport effects.

The first and for many years the only successful method for parametrization of steric effects were the E_S constants introduced by Taft [1]. These parameters were defined from chemical reactivities. They suffered from a severe deficiency in that values were available only for alkyl substituents, groups derived from these, and hydrogen. No values were extant for such groups as halogen, OX, SX, NX^1X^2, SiX_3, NMe_3^+, and PH_2. The $E_{S,\,ortho}$ values proposed by Taft were available only for a very few groups and lacked a value for hydrogen. A study of steric effects for a wide range of substituent type was therefore impossible. It was obvious at this point that further progress in the quantitative treatment of steric effects required a different approach. Such an approach was proposed by Charton who suggested that it be based upon the van der Waals radii, which have long been considered to be some kind of measure of atomic "size".

In this work we intent to consider the definition and use of steric parameters based on the van der Waals radii, although for some types of groups they have been evaluated from chemical reactivities. The methods for the estimation of steric parameters will also be presented.

1.2 The Nature of Steric Effects

Steric effects may arise in a number of ways. We consider primary steric effects to result from repulsions between nonbonded atoms. Such repulsions can only result in an increase in the energy of a group of atoms. In the case of chemical reactivities, if steric repulsions are greater in the transition state (rate data) or product (equilibrium data) than in the reactant, the latter is more stable than either of the former and compared to a system free of steric effects the rate constant or equilibrium constant will show steric diminution. If the reverse situation obtains, and steric repulsions are greater in the reactant than in the transition state or product, the former is less stable than the latter, and compared to a system free of steric effects the rate or equilibrium constant will show steric augmentation.

Secondary steric effects on chemical reactivity include:

1) A change in the solvation of the active site (group of atoms at which the observed phenomenon occurs) due to steric effects exerted by an adjacent group.

2) Variation in the concentration of a reacting conformer due to the steric effect exerted by a substituent.

3) Shielding of the active site from attack by a bulky reagent.

4) Variation in the delocalization (resonance) electrical effect of a π-bonded substituent X or active site Y attached to a π-bonded skeletal group, G. This results from twisting around the X—G or Y—G bond as a result of steric effects exerted by an adjacent group.

Physical properties depend upon the same types of steric effects as chemical reactivities. In both types of data the measurable phenomenon is occurring at a clearly defined active site. Thus, for the ionization of a set of phenols the active site is the OH group, while for the stretching frequency in the infrared spectrum of the acetyl group in a set of acetophenones the active site is the carbonyl group. In the case of bioactivities, steric effects in the formation of the bioactive substance-receptor site complex are frequently of great importance. In this case the entire bioactive substance is the active site. The same type of steric effects observed in studies of the variation of physical properties and chemical reactivity with structure are found in biological activity studies, and they can be treated in the same way.

2 Steric Parameters

2.1 Van der Waals Radii

We have noted that Van der Waals radii, r_V, have long been considered a measure of atomic size, and were so employed by Pauling[2] in his classic monograph, "The Nature of the Chemical Bond". The best extant values of r_V for atoms are those of Bondi[3,4]. These values were obtained from a careful comparison of various types of physical properties and are quite reliable. Unfortunately, values for groups are limited to the methyl group and to the half thickness of the benzene ring. If Van der Waals radii are to be used as a measure of the substituent steric effect it is of course necessary to have such values for groups as well as for atoms: Charton[5,6] has derived equations for calculating values of Van der Waals radii for symmetric tetrahedral groups, MZ^3, in which M is sp^3 hybridized.

We define the group axis of the MZ^3 group as the extension of the MG bond, where G is the skeletal group to which the MZ^3 group is attached. There are four quantities of interest for symmetrical top MZ^3 groups:

1) $r_{V,mn}$; the minimum Van der Waals radius perpendicular to the group axis (see Fig. 1)
2) $r_{V,mx}$; the maximum Van der Waals radius perpendicular to the group axis (see Fig. 1 and 2)
3) $r_{V,ax}$; the Van der Waals radius parallel to the group axis.
4) l; the covalent radius of the MZ^3 group.

These quantities may be calculated from the equations:

$$r_{V,mx} = r_{V,Z} + l_{MZ} \sin\eta \tag{1}$$

$$r_{V,mn} = r_{V,Z} + (l_{MZ} \sin\eta)/2 \tag{2}$$

$$r_{V,ax} = r_{V,Z} + l_{MZ} \cos\eta \tag{3}$$

$$l = l_{MG} + l_{MZ} \cos\eta \tag{4}$$

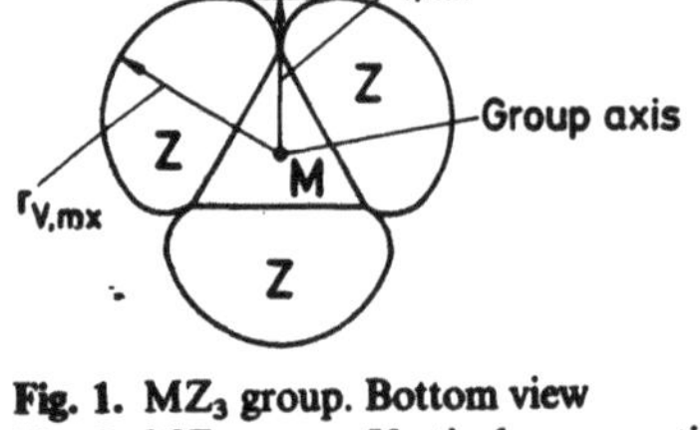

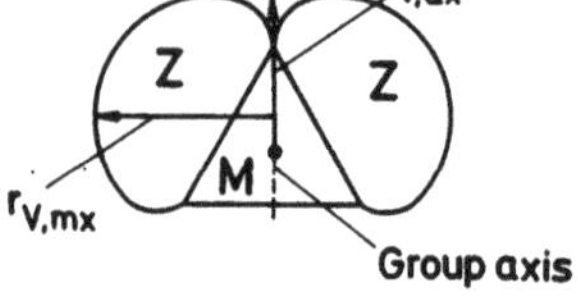

Fig. 1. MZ_3 group. Bottom view
Fig. 2. MZ_3 group. Vertical cross section

where l_{MZ} is the MZ bond length; l_{MG_i} is the MG_i bond length; $r_{V,Z}$ is the Van der Waals radius of Z; $\eta = 180° - \theta$; θ is the ZMG_i bond angle; G_i is the atom of the skeletal group G to which the MZ^3 group is bonded.

The derivation of equations 1 through 4 is given in Appendix 1. Note however that,

$$r_{V,mx} = r_{V,mn} + (l_{MZ} \sin\eta)/2 \tag{5}$$

Thus, $r_{V,mx}$ and $r_{V,mn}$ are not independent of each other. Values of $r_{V,mx}$; $r_{V,mn}$; $r_{V,ax}$; and l for various MZ^3 groups are presented in Table 2.

This method is also applicable to the calculation of these quantities for symmetric trigonal bipyramidal groups MX_4 in which M is hybridized sp^3d and to symmetric octahedral groups MX_5 in which M is hybridized sp^3d^2 as well. For the MX_4 groups (Fig. 3, 4)

$$r_{V,mx,MZ_4} = l_{MZ_e} + r_{VZ} \tag{6}$$

$$r_{V,mn,MZ_4} = r_{V,mx}/2 \tag{7}$$

$$r_{V,ax,MZ_4} = l_{MZ_a} + r_{VZ} \tag{8}$$

where l_{MZ_e} is the equatorial MZ bond length and l_{MZ_a} is the apical MZ bond length.

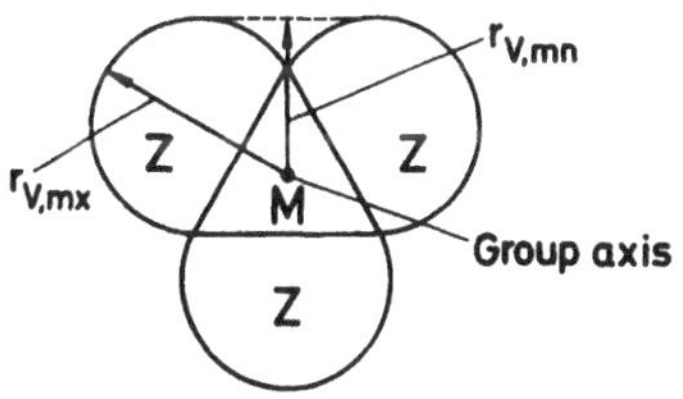

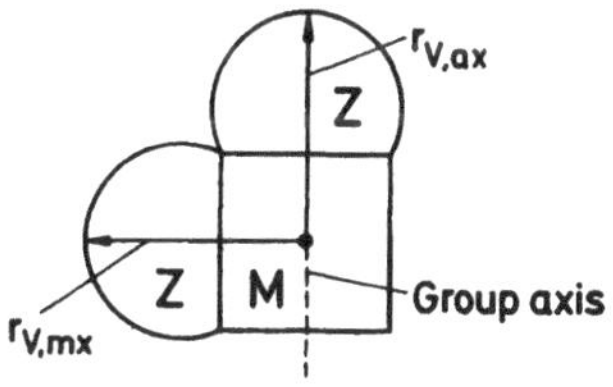

Fig. 3. MZ_4 group. Horizontal cross section through the MZ_3 plane
Fig. 4. Vertical cross section

For MZ_5 groups (Fig. 5, 6)

$$r_{V,mx,MZ_5} = l_{MZ} + r_{VZ} = r_{V,ax} \tag{9}$$

$$r_{V,mn,MZ_5} = r_{V,mx} \cos 45° \tag{10}$$

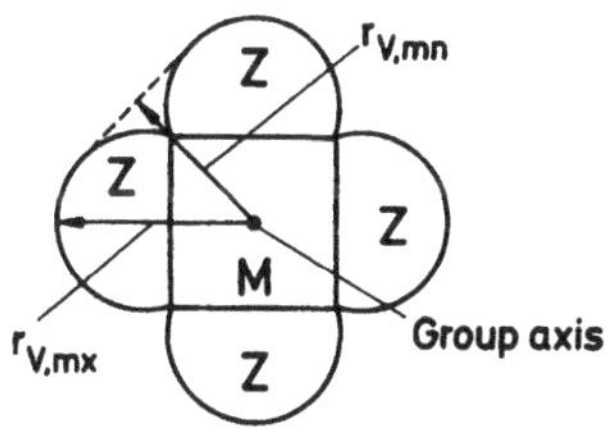

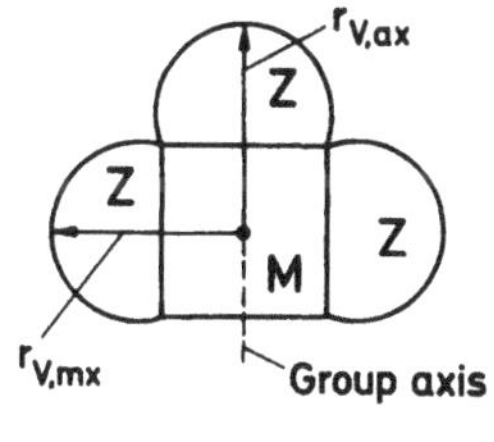

Fig. 5. MZ_5 group. Horizontal cross section through the MZ_4 plane
Fig. 6. MZ_5 group. Vertical cross section through the MZ_3 plane

Values for $r_{V,ax}$ for MX_5 are also given by Eq. 8. In Tables 1 (single atoms) and 2 (groups), the values of r_{VX} are given.

Allinger [7] has critized the Van der Waals radii of Pauling and Bondi. He argues that the Van der Waals radius must be larger than is expected from the distance between nonbonded atoms in a crystal as London forces will result in "interpenetration" of the Van der Waals radii. Allinger recommends a set of radii which are reported in Table 1. We find that the Van der Waals radii of Allinger, r_{VA}, are highly linear in those of Bondi, r_{VB}. The results of correlation with the equation

$$r_{VA,X} = a_1 r_{V,BX} + a_0 \tag{11}$$

are given in Table 3. The set includes H. F, Cl, Br, I, O, $\dot{S}$, Se, N, P, and As (set 1A). Exclusion of the value for X = H given even better results (set 1B). Allinger's radii were introduced for the purpose of carrying out force field calculations.

Bartell [8] proposed a set of one angle radii, r_0, which represent the minimum distance of approach of two atoms X and Y bonded to a central atom, M. These radii are

Table 1. Van der Waals Radii of Atoms (in Å)

X	$r_{V,B}$	$r_{V,A}$	$r_{V,G}$	X	$r_{V,B}$	$r_{V,A}$	$r_{V,G}$
A	1.88	1.92		K	(2.75)		
Al	(2.51)		1.66	Kr	2.02	2.07	
Ag	(1.72)			Li	(1.82)		
Ar	1.88			Mg	(1.73)		
As	1.85	2.20	1.61	N	1.55	1.70	1.14
Au	(1.66)			Na	(2.27)		
B	(2.13)	1.80	1.33	Ne	1.54	1.54	
Be			1.39	Ni	(1.63)		
Bi	1.87[a]		1.96	O	1.52	1.65	1.13
Br	1.85	2.10	1.59	P	1.80	2.05	1.45
C	1.70	1.85[b] 1.75[c]	1.25	Pb	(2.02)		
Cd	1.62			Pd	(1.63)		
Cu	(1.43)			Pt	(1.75)		
Cl	1.75	1.95	1.44	S	1.80	2.00	1.45
F	1.47	1.60	1.08	Sb	1.90[a]		1.88
Ga	(2.51)		1.72	Se	1.90	2.15	1.58
Ge	2.19		1.58	Si	2.10	2.10	1.55
H	1.20	1.50		Sn	(2.27)	2.40	1.88
He	1.40	1.48		Te	2.06		1.87
Hg	1.70		1.76	Tl	(1.96)		
I	1.98	2.25	1.86	U	(1.86)		
In	(2.55)		1.95	Xe	2.16		1.79
				Zn	1.39		

$r_{V,B}$ Values are those of Bondi and are from Ref. [8] and [9] unless otherwise noted. Values in parentheses are uncertain.
$r_{V,A}$ values are those of Allinger and are from Ref. [7].
$r_{V,G}$ values are nonbonded raddi of Glidewell and are from Ref. [10].
a. From Ref. [11]. b. For sp^2 and sp hybridized carbon. c. For sp^3 hybridized carbon.

Table 2. Values of $r_{V,mx}$; $r_{V,mn}$; $r_{V,ax}$; and l. Calculated from Eq. 1–4

M	Z	r_{VZ}	l_{MZ}	l_{MG_i}	$r_{V,mx}$	$r_{V,mn}$	$r_{V,ax}$	l
C	H	1.20	1.094	1.537	2.23	1.72	1.57	1.90
C	F	1.47	1.333	1.537	2.73	2.10	1.91	1.98
C	Cl	1.75	1.767	1.537	3.42	2.58	2.34	2.13
C	Br	1.85	1.938	1.537	3.68	2.76	2.50	2.18
C	I	1.98	2.139	1.537	4.00	2.99	2.69	2.25
C	Me	1.72	1.537	1.537	3.17	2.44	2.23	2.05
C	O	1.52	1.426	1.537	2.86	2.19	2.00	2.01
C	S	1.80	1.817	1.537	3.51	2.66	2.41	2.14
N	H	1.20	1.032	1.479	2.17	1.69	1.54	1.82
N	Me	1.72	1.479	1.479	3.11	2.42	2.21	1.97
Si	H	1.20	1.48	1.87	2.60	1.90	1.69	2.36
Si	F	1.47	1.57	1.87	2.95	2.21	1.99	2.39
Si	Cl	1.75	2.01	1.87	3.64	2.70	2.42	2.54
Si	Br	1.85	2.21	1.87	3.93	2.89	2.59	2.61
Si	I	1.98	2.43	1.87	4.27	3.13	2.79	2.68
Si	Me	1.72	1.87	1.87	3.48	2.60	2.34	2.49
P	H	1.20	1.42	1.841	2.54	1.87	1.67	2.32
P	O	1.52	1.537	1.841	2.97	2.24	2.03	2.35
P	Me	1.72	1.841	1.841	3.46	2.59	2.33	2.46
S	O	1.52	1.50	1.80	2.93	2.23	2.02	2.30
Ge	H	1.20	1.53	1.945	2.64	1.92	1.71	2.46
Ge	F	1.47	1.67	1.945	3.04	2.05	2.03	2.50
Ge	Cl	1.75	2.08	1.945	3.71	2.73	2.44	2.64
Ge	Br	1.85	2.29	1.945	4.01	2.93	2.61	2.71
Ge	I	1.98	2.50	1.945	4.34	3.16	2.81	2.78
Ge	Me	1.72	1.945	1.945	3.55	2.64	2.37	2.59
As	H	1.20	1.519	1.96	2.63	1.92	1.71	2.47
As	O	1.52	1.78	1.96	3.20	2.36	2.11	2.55
As	Me	1.72	1.96	1.96	3.57	2.64	2.37	2.61

Values of r_{VZ} are from the compilation of A. Bondi, J. Phys. Chem. 68, 441 (1964), except for $r_{V,Me}$ which is taken as $r_{V,mn}$ for CH_3. Values of $l_{M,Z}$ and l_{MG_i} are taken from. "Tables of Interatomic Distances", L. E. Sutton, ed., Special Publication No. 18. The Chemical Society, London, 1965.

intermediate between covalent radii and Van der Waals radii. Their values and use were described and extended by Glidewell [9,10]. They too are a linear function of r_{VB} and fit the equation

$$r_{V,GX} = a_1 r_{V,BX} + a_0 \tag{12}$$

as was shown by Charton [11]. Values of r_{VG} are also given in Table 1.

Pauling [2] proposed that ionic radii, r_I, are roughly equivalent to Van der Waals radii and can be used as estimates of the latter. Thus,

$$r_{I,X} = r_{V,X} \tag{13}$$

Correlation with the equation

$$r_{I,X} = a_1 r_{V,X} + a_0 \tag{14}$$

Table 3. Results of Correlations with Equation 11

Set	a_1	a_0	r^a	F^b	$S_{est}{}^c$	$S_{a_1}{}^c$	$S_{a_0}{}^c$	$100R^{2\ d}$	n^f	ψ^e
1A	1.09	0.0645	0.9670	129.8	0,0708	0,0961	0.164	93.52	11	0.282
1B	1.34	—0.376	0.9874	310.6	0.0396	0.0760	0.133	97.49	10	0.177
2	1.00	—0.349	0.9814	157.1	0.0488	0.0798	0.131	96.32	8	0.221
3	1.69	—1.16	0.9997	3178.0	0.0113	0.0301	0.0534^g	99.94	4	0.0356
4	1.47	—0.874	0.9880^h	81.71^i	0.0637	0.162^h	0.297^k	97.61	4	0.219
5	1.94	—1.33	0.9777^i	43.26^i	0.0796	0.296^k	0.526^l	95.58	4	0.297
6	0.254	2.17	0.9952^k	104.2^k	0.00918	0.0249^k	0.0500^h	99.05	3	0.169
7	0.699	—0.602	0.9991^j	529.3^j	0.0200	0.0304^j	0.0703^k	99.81	3	0.0752
8	1.13	—0.830	0.9936^k	77.13^k	0.0292	0.128^k	0.202^l	98.72	3	0.196
9	1.24	—1.14	0.9903	406.5	0.0433	0.0613	0.104	98.07	10	0.155

a. Correlation coefficient. b. F test for significance of the correlation. c. Standard errors of the estimate, a_1, a_0. d. Percent of the variance of the data accounted for by the correlation equation. e. Exner corrected standard error. f. Number of data points in the set. g. 99.0% Confidence level(CL). h. 98.0% C.L. i. 97.5% CL. j. 95.0% CL. k. 90.0% CL. l. 80.0% CL. Superscripts on values of r and F, indicate the confidence level, on S they indicate the confidence level of the "Student t" test for the significance of the regression coefficients. In the absence of a superscript, the confidence level is 99.9%.

shows that best results are obtained when the elements X are all members of the same group in the periodic chart and the ions all have the same charge. Results are given in Table 3 for the $C(np^2)$, $N(np^3)$, $O(np^4)$, $F(np^5)$, $Li(ns^1)$ and $Zn(nd^{10})$ groups (sets 3, 4, 5, 6, 7 and 8 respectively). As the a_1 and a_0 values obtained for different groups of elements are significantly different from each other, r_I values of elements from different groups are not directly comparable to each other.

Finally, covalent radii are a linear function of Van der Waals radii, as suggested by Pauling [2)]. Results of a correlation with the equation

$$r_{CX} = a_1 r_{VX} + a_0 \tag{15}$$

are also given in Table 3.

From the results presented above, we may conclude that:

1) The effective "size" of an atom or group varies with the phenomenon studied. Atoms are better pictured as somewhat compressible than as hard balls.
2) The atomic radius applicable to a given phenomenon is linear in the atomic radii applicable to other phenomena.

It follows, than, that for any phenomenon which is dependent on atomic or group size that there will be a radius r which is related to a fundamental or common radius, r^*, by the linear equation

$$r = a_1 r^* + a_0 \tag{16}$$

This fundamental radius, r^*, is an intensive property of the atom or group and a measure of its size in any situation. Since all of the different r values for various phenomena are a linear function of r^*, we do not need to determine r^*. We can use any r as a measure of r^*. As values of r_{VB} are available for more atoms and groups than those of any other radius, and as the values are more reliable, we choose to use r_{VB} as the standard measure of the effective size of an atom or group of atoms. We will refer to these values from now on simply as r_V values.

2.2 Requirements For A Steric Parameter in Correlation Analysis

There are certain fundamental requirements which a set of steric parameters must meet if they are to be useful in correlation analysis. These requirements include the following:

1) The steric parameters must be a measure of the "size" of the group. It is vital to distinguish clearly what is meant by size. Let us consider the steric effect of a set of alkyl groups C_nH_{2n+1} with n held constant and greater than 3. We find that the steric effect exerted varies considerably with the degree of branching. Thus, for example when n = 7, Hp has about one third the steric effect of CEt_3. Obviously the steric effect varies considerably with the degree of branching. The volume of all alkyl groups with the same value of n must be constant, as the polarizability is proportional to the number of carbon atoms in the group, and the group volume is a linear function of group polarizability. Clearly, the steric effect of a group is not dependent upon its volume. By group size, then, we actually mean the distance from the center of the group to its perimeter. For a nonsymmetric group there is more than one such distance. In most examples of steric effects exerted upon chemical reactivity or a physical property, the group X interacts with an active site Y at which some measurable property takes place. Only a single distance from group center to perimeter is of interest. It is readily apparent that the steric effect is a directed quantity, that is, it is a vector rather than a scalar. Volume is a scalar quantity.
2) It is of the *utmost* importance that steric parameters must be either experimentally measurable or calculable for *all* possible types of substituent.
3) The steric parameter must be as nearly independent as possible of electrical and transport effects. It must of course be realized that a small subset of badly chosen groups will frequently show a statistically significant relationship between steric and electrical parameters or between steric and transport parameters. Often this can ge avoided by proper experimental design. In some cases the nature of the problem makes it impossible to achieve independence of steric parameters. Correlations of amino acid bioactivities which are restricted to the amino acids that are common constituents of protein are an example of a data set whose members cannot be conveniently varied.
4) Experimental data from which steric parameters are obtained should be readily, conveniently, precisely and accurately determinable.
5) Methods should be available for the estimation of steric parameters for those groups for which experimental data is unavailable.
6) In accord with other parameters used in correlation analysis the steric parameter should have a value of zero for X=H. In this way the intercept of the correlation equation will be the calculated value for X=H.

2.3 The Definition of υ

We have remarked above that the Taft E_S values suffered from a number of deficiencies. In fact, the only direct evidence that they were a measure of steric effects was their successful correlation with r_V values, a correlation limited to symmetrical top tetrahedral substituents such as CH_3 and CF_3 and to H. The evidence presented above indicates that r_V values are a useful measure of steric effects and suggested that they might be used directly as steric parameters in correlation analysis. They were so used by Charton [6,12,13,14]. They did meet the first and third criteria for steric parameters in full and the fourth in part. They did not meet the second, fifth and sixth, however. The Taft E_S values met the first criteria only if it could be assumed that the use by Taft of average values of data obtained under different experimental conditions was valid. Our results indicate that this is not the case. E_S values did meet the third condition, but were unable to meet the other criteria. It seemed more reasonable to base a set of steric parameters on the van der Waals radii than to do so upon E_S values and to attempt to remove their deficiencies. By defining a set of steric parameters, designated υ values, from the equation

$$\upsilon_X \equiv r_{VX} - r_{VH} = r_{VX} - 1.20 \tag{17}$$

using the Bondi scale, the sixth criterion is met.

3 The Dependence of Steric Effects on Conformation

3.1 Substituent Classification

The steric effect of most groups is conformationally dependent to some degree. In order to parameterize steric effects properly we must classify groups according to the extent to which their steric effects are conformationally dependent. Our first category is that of groups which have no conformational dependence (NCD groups). Such groups are:

Ia. Single atom substituents — examples are H, F, Cl, Br, I, O^-, S^-

Ib. Cylindrical substituents — examples are —CN, —NC, —C≡CZ where Z=H, F, Cl, Br. I.

The second class is that of groups which exhibit a minimum conformational dependence (MCD groups). They include:

IIa. Symmetric tetrahedral groups of the type MZ_3. Examples are CH_3, CF_3, $SiCl_3$, GeH_3, $SnMe_3$

IIb. Symmetric bicycloalkyl groups. An example is the 1-bicyclo[2.2.2]octyl group.

IIc. Symmetric trigonal bipyramidal groups of the type MZ_4. No common examples are known, the only known example is PF_4.

IId. Symmetric octahedral groups of the type MZ_5. The only common example is SF_5.

IIe. Groups of the type $M(lp)_nH_{3-n}$ where n = 1 or 2 and lp represents a lone pair. Examples are OH, SH, SeH, NH_2, PH_2.

The third category is that of groups with an intermediate degree of conformational dependence of their steric effects (ICD groups). Included in this class are:

IIIa. Tetrahedral groups of the type $MZ^1_2Z^2$ ($Z^1 \neq Z^2$). Examples are CF_2H, $SiMe_2H$, CCl_2H

IIIb. Cycloalkyl groups with four or more ring atoms and their heterocyclic analogs. Examples are cyclohexyl, 4-piperidinyl, and 3-tetrahydrofuryl.

IIIc. Groups of the type $M(lp)_nZ_{3-n}$. Examples are OEt, OPh, StBu, NHMe, NEt_2.

The final category is that of groups which show a strong dependence of their steric effect upon conformation (SCD groups). Among the members of this class are:

IVa. Tetrahedral groups of the type $MZ^1Z^2Z^3$ ($Z^1 \neq Z^2 \neq Z^3$). Examples are CHIMe, SiFBrMe, CMe, iPr, Bu

IVb. Planar π bonded groups. Examples are NO_2, Ph, Ac, CO_2Me, CHO and $CONH_2$.

IVc. Cyclopropyl and its heterocyclic analogs.

3.2 Steric Parameters of NCD and MCD Groups

The minimal set of parameters required to describe the steric effect of a group are based on the $r_{V,mx}$; $r_{V,mn}$ and $r_{V,ax}$ values. The $r_{V,mx}$ value is the maximum Van der Waals radius perpendicular to the group axis. The group axis, which is defined in Fig. 1, is the extension of the X—G bond which joins the substituent X to the skeletal group G. The $r_{V,mn}$ value is the minimum Van der Waals radius perpendicular to the group axis. The $r_{V,ax}$ value is the Van der Waals radius parallel to the group axis. Thus, $r_{V,ax}$ can be thought of as the length of the group, $r_{V,mx}$ its maximum width and $r_{V,mn}$ its minimum width.

There are, then, three values, υ_{mx}, υ_{mn} and υ_{ax} which are the minimum necessary to characterize most substituents. For type Ia (monatomic) substituents $\upsilon_{mx} = \upsilon_{mn} = \upsilon_{ax}$. For type Ib (cylindrical) substituents, $\upsilon_{mx} = \upsilon_{mn}$. For all other groups, all three values are usually different. Values of υ for monatomic groups are easily obtainable from Eq. 17 using the r_V values given in Table 1. All υ values are reported in Table 4. Values of υ_{mn} and υ_{mx} for cylindrical groups can be obtained (Fig. 7) using the r_V values given by Bondi[3] for $C{\equiv}C$ (1.78) and $C{\equiv}N$ (1.60). The υ_{ax} value for any triply bonded group ($-M^1{\equiv}M^2Z$) is given by

$$\upsilon_{ax} = l_{M^1M^2} + l_{M^2Z} + r_{V,Z} - 1.20 \tag{18}$$

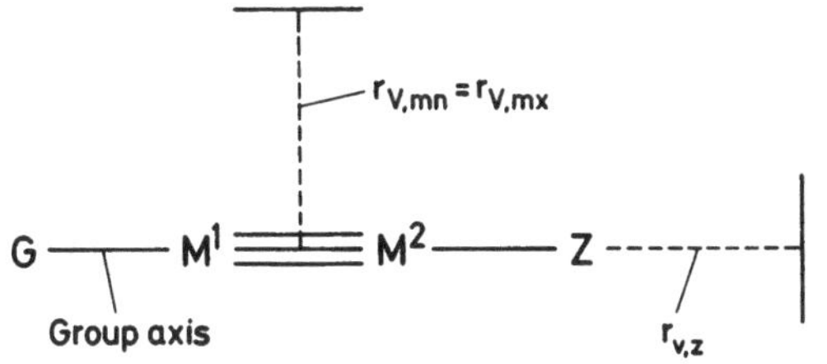

Fig. 7. Cylindrical triply bonded group

where $l_{M^1M^2}$ and l_{M^2Z} are the bond lengths of the M^1M^2 and M^2Z bonds respectively, and $r_{V,Z}$ is the $r_{V,ax}$ value for the Z group. For a triply bonded group $—M^1 \equiv M^2$, υ_{ax} is given by

$$\upsilon_{ax} = l_{M^1M^2} + r_{V,ax,M^2} \tag{19}$$

Values of υ_{mn}, υ_{mx} and υ_{ax} for symmetrical tetrahedral (MZ_3), trigonal bipyramidal (MZ_4), and octahedral (MZ_5) groups can be obtained from Eq. 17 using the r_V values calculated from Eq. 1–3 and 6–10. Values of $r_{V,mn}$, $r_{V,mx}$ and $r_{V,ax}$ for symmetrical bicycloalkyl and heterobicycloalkyl groups can be calculated from molecular geometry. The appropriate υ values are then obtainable from Eq. 17.

Our results support the argument that $r_{V,mn}$ for an $M(lp)_nH_{3-n}$ group is equal to r_V for the atom M. $r_{V,mx}$ and $r_{V,ax}$ can be calculated as shown in Fig. 8, using the equations:

$$r_{V,mx} = l_{MH} \sin \eta + r_{V,H} \tag{20}$$

$$r_{V,ax} = l_{MH} \cos \eta + r_{V,H} \tag{21}$$

with

$$\eta = (180° - \sphericalangle H—M—G) \tag{22}$$

Appropriate υ values can now be calculated from Eq. 17.

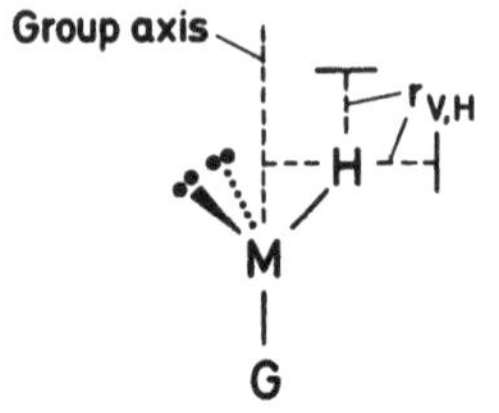

Fig. 8. $M(1p)_nH_{3-n}$ group. The example shown is $M(1p)_2H$

Table 4. υ Values

X	Class	υ_{mn}	υ_{mx}	υ_{ax}	υ_{ef}	Ref.
Alkyl Groups						
Me	M	0.52	1.03	0.37	0.52	a, b
Et	I				0.56	b
Pr	I				0.68	b
iPr	I				0.76	b
Bu	I				0.68	b
iBu	I				0.98	b
sBu	I				1.02	b
tBu	M	1.24	1.97	1.03	1.24	a, b
Am	I				0.68	b
$sBuCH_2$	I				1.00	b
$iPrCH_2CH_2$	I				0.68	b

Table 4. (continued)

X	Class	υ_{mn}	υ_{mx}	υ_{ax}	υ_e	Ref.
$tBuCH_2$	I				1.34	b
Et_2CH	I				1.51	b
MePrCH	I				1.05	c
MeiPrCH	I				1.29	c
Me_2EtC	I				1.63	b
Hx	I				0.73	b
$tBuCH_2CH_2$	I				0.70	b
EtPrCH	I				1.51	c
MeBuCH	I				1.07	c
MeiBuCH	I				1.09	c
$iPr(CH_2)_3$	I				0.68	c
tBuCHMe	I				2.11	b
iPrCHEt	I				2:11	b
Hp	I				0.73	b
Pr_2CH	I				1.54	b
EtBuCH	I				1.55	b
$tBuCH_2CHMe$	I				1.41	b
$tBuCMe_2$	I				2.43	b
Et_3C	I				2.38	b
Oc	I				0.68	b
$tBuCH_2CMe_2$	I				1.74	b
No	I				0.68	c
Bu_2CH	I				1.56	c
$(iPrCH_2)_2CH$	I				1.70	c
$tBuCHEtCH_2CH_2$	I				1.01	b
Ud	I				0.68	c
$(tBuCH_2)_2CH$	I				2.03	b
Td	I				0.68	c
Pnd	I				0.68	c
Hpd	I				0.68	c
Cycloalkyl						
cPr	L				1.06	c
cBu	I				0.51	c
cAm	I				0.71	c
cHx	I				0.87	b
cHp	I				1.00	c
$cHxCH_2$	I				0.97	b
$cHx(CH_2)_2$	I				0.70	b
$cHx(CH_2)_3$	I				0.71	b
1-bicyclo[2.2.2]octyl	M				1.33	d
1-adamantyl	M				1.33	d
Vinylalkyl, Arylalkyl						
$ViCH_2$	I				0.69	c
$Vi(CH_2)_2$	I				0.75	c
$Vi(CH_2)_3$	I				0.75	c
$Me(ViCH_2)CH$	I				1.04	c
$PhCH_2$	I				0.70	b
$PhCH_2CH_2$	I				0.70	b
PhMeCH	I				0.99	b

Table 4. (continued)

X	Class	υ_{mn}	υ_{mx}	υ_{ax}	υ_{ef}	Ref.
$Ph(CH_2)_3$	I				0.70	b
$Me(PhCH_2)CH$	I				0.98	c
PhEtCH	I				1.18	b
$Ph(CH_2)_4$	I				0.70	b
9-fluorenyl	I				1.08	c
Ph_2CH	I				1.25	b
9-methyl-9-fluorenyl	I				1.41	c
Ph_2CMe	I				2.34	c
9-ethyl-9-fluorenyl	I				1.53	c
Ph_2CEt	I				2.75	c
9-isopropyl-9-fluorenyl	I				2.21	c
$ViPh_2CCH_2$	I				2.74	c
9-tBu-9-fluorenyl	I				2.63	c
9-Ph-9-fluorenyl	I				1.59	c
Ph_3C	I				2.92	c
$9\text{-}PhCH_2\text{-}9$-fluorenyl	I				1.63	c
$Z\text{-}(CH_2)_{11}CH{=}CHOc$	I				0.67	c
$E\text{-}(CH_2)_{11}CH{=}CHOc$	I				0.68	c
Haloalkyl						
CH_2F	I				0.62	b
CH_2Cl	I				0.60	b
CH_2Br	I				0.64	b
CH_2I	I				0.67	b
CHF_2	I				0.68	b
$CHCl_2$	I				0.81	b
$CHBr_2$	I				0.89	b
CHI_2	I				0.97	b
CF_3	M	0.90	1.53	0.71	0.90	a, b
CCl_3	M	1.38	2.22	1.14	1.38	a, b
CBr_3	M	1.56	2.48	1.30	1.56	a, b
CI_3	M	1.79	2.80	1.49	1.79	a, b
CH_2CH_2Cl	I				0.97	c
CH_2CH_2Br	I				0.92	d
CH_2CH_2I	I				0.93	c
$MeBr_2C$	I				1.46	b
CHBrMe	I				0.75	d
CHClMe	I				0.70	d
Me_2CBr	I				1.39	b
Oxaalkyl						
$MeOCH_2$	I				0.63	c
$EtOCH_2$	I				0.61	c
$MeOCH_2CH_2$	I				0.89	c
$MeOCH_2OCH_2$	I				0.62	c
$PrOCH_2$	I				0.65	c
$iPrOCH_2$	I				0.67	c
$EtOCH_2CH_2$	I				0.89	c
$MeO(CH_2)_3$	I				0.69	c
MeCHOEt	I				0.75	d
EtCHOMe	I				1.22	c

Table 4. (continued)

X	Class	υ_{mn}	υ_{mx}	υ_{ax}	υ_{ef}	Ref.
$MeOCH_2CH_2OCH_2$	I				0.57	c
$BuOCH_2$	I				0.66	c
$iBuOCH_2$	I				0.62	c
$PrOCH_2CH_2$	I				0.89	c
$iPrOCH_2CH_2$	I				0.87	c
$EtO(CH_2)_3$	I				0.69	c
$MeO(CH_2)_4$	I				0.68	c
PrCHOMe	I				1.22	c
$EtOCH_2CH_2OCH_2$	I				0.56	c
$MeOCH_2CH_2OCHMe$	I				0.67	c
$BuOCH_2CH_2$	I				0.89	c
$iBuOCH_2CH_2$	I				0.89	c
$PrO(CH_2)_3$	I				0.70	c
$EtO(CH_2)_4$	I				0.67	c
BuCHOMe	I				1.20	c
$PrOCH_2CH_2OCH_2$	I				0.56	c
$MeOCH_2CH_2OCH_2CH_2OCH_2$	I				0.56	c
$BuO(CH_2)_3$	I				0.71	c
$BuOCH_2CH_2OCH_2$	I				0.55	c
$tBuOOCMe_2$	I				1.49	c
$tBuOCH_2CMe_2$	I				1.30	c
$tBuCH_2OCMe_2$	I				1.23	c
Hydroxy alkyl						
CH_2OH	I				0.53	c
CO_3	M	0.99	1.66	0.80	0.99	a, b
CH_2CH_2OH	I				0.77	c
MeCHOH	I				0.50	c
EtCHOH	I				0.71	c
$CH_2CHOHMe$	I				0.86	d
CMe_2OH	I				0.76	d
PrCHOH	I				0.71	c
CH_2CMe_2OH	I				1.06	d
BuCHOH	I				0.70	c
$CHMeCMe_2OH$	I				1.45	d
$CH_2CMeEtOH$	I				1.13	d
CHMeCMeEtOH	I				1.48	d
$CHEtCMe_2OH$	I				1.64	d
Aminoalkyl						
CH_2NH_2	I				0.54	e
$MeCHNH_2$	I				0.58	e
$EtCHNH_2$	I				0.89	e
$PrCHNH_2$	I				0.89	e
$iPrCHNH_2$	I				1.38	e
$BuCHNH_2$	I				0.91	e
$iBuCHNH_2$	I				0.85	e
$sBuCHNH_2$	I				1.42	e

Table 4. (continued)

X	Class	υ_{mn}	υ_{mx}	υ_{ax}	υ_{ef}	Ref.
Thia alkyl groups						
CS_3	M	1.46	2.31	1.21	1.46	a, b
$MeSCH_2$	I				0.70	c
$EtSCH_2$	I				0.71	c
$MeSCH_2CH_2$	I				0.78	c
2,5-dithiacyclopentyl	I				0.89	c
$EtSCH_2CH_2$	I				0.79	c
EtSCHMe	I				1.10	c
2,5-dithiacyclohexyl	I				1.16	c
$(EtS)_2CH$	I				1.39	c
Alkoxy Groups						
OMe	I				0.36	g
OEt	I				0.48	g
OPr	I				0.56	g
OiPr	I				0.75	g
OBu	I				0.58	g
OiBu	I				0.62	g
OsBu	I				0.86	g
OtBu	I				1.22	g
OAm	I				0.58	g
OCH_2sBu	I				0.62	g
OCH_2iBu	I				0.62	g
OCH_2tBu	I				0.70	g
$OCHEt_2$	I				1.00	g
OCHMePr	I				0.90	g
OCHMeiPr	I				0.91	g
$OCMe_2Et$	I				1.35	g
OHx	I				0.61	g
OCHEtPr	I				1.04	g
OCH_2CH_2tBu	I				0.53	g
OCH_2CHEt_2	I				0.71	g
OCH_2CMe_2Et	I				0.78	g
$OCH_2CHMeiPr$	I				0.64	g
OCHEtiPr	I				1.18	g
$OCEt_2Me$	I				1.52	g
$OCPrMe_2$	I				1.39	g
OCHMetBu	I				1.19	g
OCHMeAm	I				0.90	g
$OCH_2CHEtiPr$	I				0.76	g
$OCH_2CHMetBu$	I				0.66	g
OCH_2CMeEt_2	I				0.82	g
OOc	I				0.61	g
$OCH_2CHEtBu$	I				0.76	g
$OCH_2CHEt(tBu)$	I				0.96	g
OCHMeHx	I				0.92	g
OCH_2CHiPr_2	I				0.89	g
$OCHCEt_3$	I				0.97	g
$OCHiBu_2$	I				1.28	g
ODod	I				0.65	g

Table 4. (continued)

X	Class	υ_{mn}	υ_{mx}	υ_{ax}	υ_{ef}	Ref.
Cycloalkoxy Groups						
OCH_2cPr	I				0.48	g
OcAm	I				0.77	g
OCH_2cBu	I				0.52	g
OcHx	I				0.81	g
OCH_2cAm	I				0.58	g
OCH_2cHx	I				0.65	g
Other OX Groups						
OCH_2CH_2OH	I				0.46	f
OCH_2CH_2Cl	I				0.51	f
OCH_2CH_2Br	I				0.58	f
$O(CH_2)_3Cl$	I				0.52	f
$O(CH_2)_4OMe$	I				0.54	f
Alkylamino, Cyclo Alkylamino, and other NHX Groups						
NHMe	I				0.39	h
NHEt	I				0.59	h
NHPr	I				0.64	h
NHiPr	I				0.91	h
NHBu	I				0.70	h
NHiBu	I				0.77	h
NHsBu	I				1.12	h
NHtBu	I				1.83	h
NHAm	I				0.64	h
$NHCH_2iBu$	I				0.65	h
NHcHx	I				0.92	h
NHHx	I				0.66	h
$NHCH_2Ph$	I				0.62	h
Dialkylamino Groups						
NMe_2	I				0.43	h
NMeEt	I				0.87	h
NEt_2	I				1.37	h
NPr_2	I				1.60	h
$NiPr_2$	I				2.01	h
Alkyl thio Groups						
SMe	I				0.64	i
SEt	I				0.94	i
SPr	I				1.07	i
SiPr	I				1.19	i
SBu	I				1.15	i
SiBu	I				1.15	i
SsBu	I				1.36	i
StBu	I				1.60	i

Table 4. (continued)

X	Class	υ_{mn}	υ_{mx}	υ_{ax}	υ_{ef}	Ref.
Silyl, Germyl, Stannyl Groups						
SiH_3	M	0.70	1.40	0.49	0.70	a
SiF_3	M	1.01	1.75	0.79	1.01	a
$SiCl_3$	M	1.50	2.44	1.22	1.50	a
$SiBr_3$	M	1.69	2.73	1.39	1.69	a
SiI_3	M	1.93	3.07	1.59	1.93	a
$SiMe_3$	M	1.40	2.28	1.14	1.40	a
GeH_3	M	0.72	1.44	0.51	0.72	a
GeF_3	M	1.05	1.84	0.83	1.05	a
$GeCl_3$	M	1.53	2.51	1.24	1.53	a
$GeBr_3$	M	1.73	2.81	1.41	1.73	a
GeI_3	M	1.96	3.14	1.61	1.96	a
$GeMe_3$	M	1.44	2.35	1.17	1.44	a
SnH_3	M					
SnF_3	M					
$SnCl_3$	M					
$SnBr_3$	M					
SnI_3	M					
$SnMe_3$	M	1.55	2.57	1.25	1.55	a
Ammonio, Phosphonio, Arsonio Groups						
NH_3^+	M	0.49	0.97	0.34	0.49	a
NMe_3^+	M	1.22	1.91	1.01	1.22	a
PH_3^+	M	0.67	1.34	0.47	0.67	a
PMe_3^+	M	1.39	2.26	1.13	1.39	a
AsH_3^+	M	0.72	1.43	0.51	0.72	a
$AsMe_3^+$	M	1.44	2.37	1.17	1.44	a
Ethynyl, Cyano, Isocyano and Related Groups						
—CN	N	0.40	0.40		0.40	j
—NC	N	0.40	0.40		0.40	j
—C≡CH	N	0.58	0.58		0.58	j
—C≡CMe	N	0.58	0.58		0.58	j
$—C≡CCF_3$	N	0.58	0.58		0.58	j
—C≡C—C=CH	N	0.58	0.58		0.58	j
—C≡C—C≡C—Me	N	0.58	0.58		0.58	j
—C≡C—Ph	M	0.58	0.58		0.58	j
Miscellaneous Substituted Alkyl Groups						
CH_2NO_2	I				0.86	d
$CH_2SO_3^-$	I				0.91	d
CH_2CN	I				0.89	b
CH_2Ac	I				0.76	d
CH_2CO_2Me	I				0.58	d
CH_2CH_2Ac	I				0.81	d
CH_2SiMe_3	I				1.49	e
PhCHCl	I				1.20	f
$PhOCH_2$	I				0.74	c
$PhSCH_2$	I				0.82	c

Table 4. (continued)

X	Class	υ_{mn}	υ_{mx}	υ_{ax}	υ_{ef}	Ref.
PhCHOH	I				0.69	f
9-hydroxy-9-fluorenyl	I				0.98	c
Ph_2COH	I				2.04	f
$C(CN)_3$	M	0.90	1.44	0.74	0.90	a, n
$C(C_2H)_3$	M	1.06	2.13	0.75	1.06	a, n
Miscellaneous Groups Without Carbon						
H	N	0	0	0	0	b
F	N	0.27	0.27	0.27	0.27	b
Cl	N	0.55	0.55	0.55	0.55	b
Br	N	0.65	0.65	0.65	0.65	b
I	N	0.78	0.78	0.78	0.78	b
OH	M				0.32	k
NH_2	M				0.35	k
SH	M				0.60	k
Ph_2	M				0.60	k
SeH	M				0.70	k
AsH_2	M				0.65	k
SO_3^-	M	1.03	1.73	0.82	1.03	a
SF_5	M				1.37	l
PO_3^{2-}	M	1.04	1.77	0.83	1.04	a
AsO_3^{2-}	M	1.16	2.00	0.91	1.16	a
Planar π Bonded Groups						
NO_2	L	0.35	1.39			m
CO_2H	L	0.50	1.39			m
CO_2Ak	L	0.50	1.39			m
$CONH_2$	L	0.50	1.39			m
Ac	L	0.50	1.39			m
Bz	L	0.50	1.39			m
CHO	L	0.50	1.39			m
Vi	L	0.57	0.95			m
Ph	L	0.57	2.15			m

a. Ref. [5], Table 2 and Eq. 17
b. Ref. [24]
c. Ref. [25]
d. Calculated from rates of acid catalyzed hydrolysis of XCO_2Et in H_2O at 25°
e. Calculated from the data of R. W. Hay and P. J. Morris, *J. Chem. Soc. B.* 1577 (1970), R. W. Hay and L. J. Porter, *ibid.*, 1261 (1967) using a correlation of log kr with υ_{AkCHMe} to determine S and the value of CH_2NH_2 to calculate f.
f. Ref. [68]
g. Ref. [26]
h. Ref. [27]
i. Ref. [28]
j. Ref. [32]
k. From Section 3.2
l. From Eq. 10 and Eq. 17
m. From Section 3.6
n. The gear effect may be important for thesé groups
For abbreviations, see Appendix 2. Under class, CD follows all abbreviations

3.3 The Minimal Steric Interaction Principle

We have noted (1.1) that primary steric effects involve repulsions between non-bonded atoms which raise the energy of a system. It follows, then, that a group of atoms will prefer that conformation which minimizes these repulsions. This is the principle of minimal steric interaction (MSI). It implies that the steric effect exerted by any atom or group of atoms is the smallest possible. In the simplest case this means that the steric effect of a group X is dependent upon its $\upsilon_{mn,X}$ value. Actually, the steric effect of a group in a particular system depends on the phenomenon which is studied. Thus, taking reaction rates as an example, the steric requirements of a transition state vary with the nature of the reaction. Although in every case the MSI principle applies, and a group will exert the smallest possible steric effect, the effective υ value can vary from one reaction to another. In general, however, the steric effect of an MZ_3 group is taken to be equal to υ_{mn} for the group.

3.4 The Gear Effect

A number of authors have suggested that a tetrahedral group MZ_3 behaves like a three-toothed gear and that when two such gears mesh, the effective size and therefore the steric effect of the MZ_3 group is reduced [15–23]. Consider the conformations 1 (meshed) and 2 (non-meshed) of the groups $Z^1Z^1_3$ and $M^2Z^2_3$ shown in Figs. 9A and 9B. The non-meshed conformation is the one we have previously assumed in calculating $r_{V,mn}$ for MZ_3 groups. The difference between the steric effect exerted by an MZ_3 group in conformation 1 and that due to an MZ_3 group in conformation 2 can be given by the difference between the distances $\overline{C^1{-}C^2}$ and $\overline{C^1{-}Z^2}$. Z^1 and Z^2 are the nuclei of the Z atoms in the groups $M^1Z^1_3$ and $M^2Z^2_3$. C^1 and C^2 are the centers of the line sigments joining the two Z^1 atoms an the two Z^2 atbms in contact in the non-meshed conformation and the two Z^1 atoms in contact with Z^2 in the meshed conformation. Clearly in the non-meshed conformation

$$\overline{Z^1Z^2} = \overline{C^1C^2} = r_{V,Z^1} + r_{V,Z^2} = \Sigma r_V \tag{23}$$

where r_{V,Z^1} and r_{V,Z^2} are the Van der Waals radii of Z^1 and Z^2 respectively. For an MZ_3 group with regular tetrahedral geometry $\measuredangle Z{-}M{-}Z = 109.5°$. From Fig. 10 it is clear that the line segment CM besects $\measuredangle Z{-}M{-}Z$ and therefore $\measuredangle C{-}M{-}Z = 54.75°$. Then, as $\overline{MZ}$ is the length of the M—Z bond (l_{MZ})

$$\overline{C^1Z^1} = l_{MZ} \sin 54.75° \tag{24}$$

since $Z^1CZ^2 = 90°$

$$\overline{C^1Z^2} = [(\overline{Z^1Z^2})^2 - (\overline{C^1Z^1})^2]^{1/2} \tag{25}$$

$$= [(\Sigma r)^2 - l^2_{MZ} \sin^2 54.75°]^{1/2} \tag{26}$$

We define

$$\Delta \equiv \overline{C^1C^2} - \overline{C^1Z^1} \tag{27}$$

Then the gear effect Van der Waals radius of a tetrahedral MZ_3 group is taken to be

$$r_{V,gr,MZ_3} = r_{V,mn,MZ_3} - (\Delta/2) \tag{28}$$

Values of r_{V,gr,MZ_3} and Δ are reported in Table 5.

In considering the importance of the gear effect it is convenient to examine the quantity P_{gr} defined by the expression

$$P_{gr} = \frac{(r_{V,mn,MZ_3} - r_{V,gr,MZ_3})}{r_{V,mn,MZ_3}} \cdot 100 \tag{29}$$

$$= \frac{[r_{V,mn,MZ_3} - r_{V,mn,MZ_3} + (\Delta/2)]}{r_{V,mn,MZ_3}} \cdot 100 \tag{30}$$

$$= \frac{(\Delta/2) \cdot 100}{r_{V,mn,MZ_3}} \tag{31}$$

P_{gr} represents the percent change in the magnitude of the minimum perpendicular Van der Waals radius of MZ_3 due to the gear effect. Values of P_{gr} are given in Table 5.

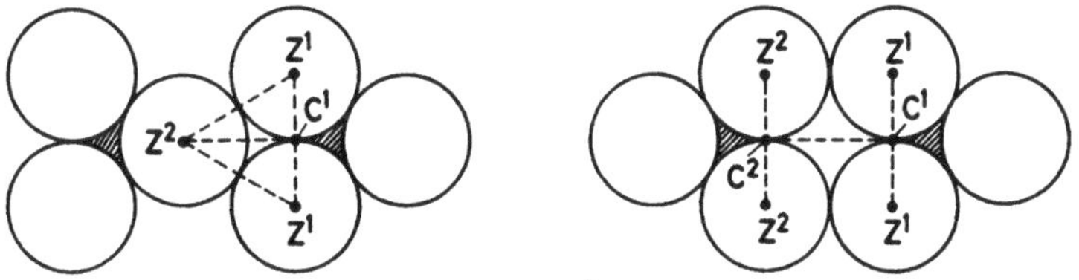

a b

Fig. 9A. Gear effect. Top view of conformation 1 (meshed)
Fig. 9B. Gear effect. Top view of conformation 2 (nonmeshed)

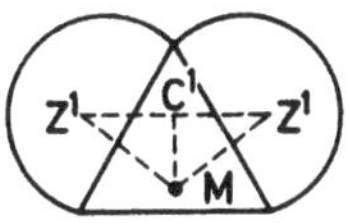

Fig. 10. Gear effect. Vertical cross section through an MZ_3 group

Many values of r_V for groups have been obtained from kinetic data where the experimental errors is probably at last 5% in log kr. It seems likely then, that the gear effect in the Me, CF_3, CMe_3, NH_3^+, and NMe_3^+ is negligible in most cases. The gear effect will clearly be much more important in MZ_3 groups with M = Si, Ge, and As.

Table 5. Values of Δ, P_{gr} and $r_{V,gr}$

M	Z	r_Z [a]	l_{MZ} [b]	$r_{V,mn}$ [c]	Δ [d]	P_{gr} [e]	$r_{V,gr}$ [f]
C	H	1.20	1.094	1.72	0.172	5.23	1.63
C	F	1.47	1.333	2.10	0.209	4.76	2.00
C	Cl	1.75	1.767	2.58	0.311	6.20	2.42
C	Br	1.85	1.938	2.76	0.356	6.52	2.58
C	I	1.98	2.139	2.99	0.406	6.69	2.79
C	Me	1.72	1.537	2.44	0.237	4.92	2.32
C	O	1.52	1.426	2.19	0.232	5.48	2.07
C	S	1.80	1.817	2.66	0.320	6.02	2.50
N	H	1.20	1.032	1.69	0.153	4.73	1.61
N	Me	1.72	1.479	2.42	0.219	4.55	2.31
Si	H	1.20	1.48	1.90	0.327	8.42	1.74
Si	F	1.47	1.57	2.21	0.294	6.79	2.06
Si	Cl	1.75	2.01	2.70	0.409	7.41	2.50
Si	Br	1.85	2.21	2.89	0.470	8.30	2.65
Si	I	1.98	2.43	3.13	0.533	8.63	2.86
Si	Me	1.72	1.87	2.34	0.358	7.69	2.16
P	H	1.20	1.42	1.87	0.299	8.02	1.72
P	O	1.52	1.537	2.24	0.271	6.25	2.10
P	Me	1.72	1.841	2.59	0.346	6.56	2.42
S	O	1.52	1.50	2.23	0.258	5.83	2.10
Ge	H	1.20	1.53	1.92	0.351	9.38	1.74
Ge	F	1.47	1.67	2.05	0.335	8.29	1.88
Ge	Cl	1.75	2.08	2.73	0.440	8.06	2.51
Ge	Br	1.85	2.29	2.93	0.507	8.53	2.68
Ge	I	1.98	2.50	3.16	0.567	8.86	2.88
Ge	Me	1.72	1.945	2.64	0.389	7.20	2.45
As	H	1.20	1.519	1.92	0.345	8.85	1.75
As	O	1.52	1.78	2.36	0.370	8.05	2.17
As	Me	1.72	1.96	2.64	0.395	7.58	2.44
Sn	Me	1.72	2.18	2.75	0.496	9.09	2.50

a. From Table 1 except for Me which takes the $r_{V,mn}$ value. b. From Ref. [7]. c. Calculated as described in Charton, M.: Prog. Phys. Org. Chem. *8*, 235 (1971). d. Calculated from Eq. 27. e. Calculated from Eq. 28. f. Calculated from Eq. 29.

3.5 Steric Parameters of ICD Groups

The dependence of the steric effect of ICD groups on conformation makes it difficult to determine υ_{mn}, υ_{mx} and υ_{ax} for many of these groups. A second problem arises with groups (1) that have a skeleton of 2 or more atoms other than H. Although υ_{mn} is frequently determined by the first atom of the skeleton, M^1, the remaining atoms also contribute to the steric effect of the group. This is not a problem with NCD or MCD groups. For these classes υ_{mn} is a valied measure of the steric effect.

$$G{-}M^1{-}M^2{-}M^3{-} \tag{1}$$

Taft [1] based his steric effect constants on the assumption that rates of esterification of carboxylic acids with alcohols and of acid catalyzed hydrolysis of carboxylate esters were structurally dependent on steric effects. This assumption was supported by the work of Charton [5], who showed that the Taft E_S values for H, an NCD group, and for the MCD groups, CH_3, CCl_3, CBr_3, CF_3, were well correlated by the equation

$$E_{SX} = a_1 r_{V,mn,X} + a_0 \tag{32}$$

In his calculation of E_S values, Taft made use of average values of log (kr_X/kr_O). Evidence has been presented which suggests that this procedure is unjustified, as the dependence of log $k_{VX}/k_{VO}0)$ on X is *not* independent of medium. Effective values, υ_{ef}, of the steric parameter for some ICD groups have been obtained by means of a two step procedure, 24, 25:

1) Those log kr values which are available for NCD and MCD groups are correlated with the equation

$$\log kr_X = S\upsilon_{mn,X} + h \tag{33}$$

Only log kr values obtained under the same reaction conditions (such as pressure, temperature and medium) are included in a data set.

2) The S and h values obtained from the correlation are used to calculate values of υ_{ef} from the equation

$$\upsilon_{ef,X} = (\log kr_X - h)/S \tag{34}$$

Values of υ_{ef} are given in Table 4. The method is useful only for ICD groups in which M^1 is carbon, as it is restricted to acyl substituted esters. It is therefore applicable to $CZ^1Z^2Z^3$ and to cycloalkyl groups.

A different procedure was required for the definition of steric parameters of OZ, SZ, and NZ^1Z^2 groups as no MCD or NCD groups were available to make possible the use of the method used above. Following the treatment of Charton [26], we may write any group X in the form M^1Z where M^1 is that atom of the group X which is bonded to the skeletal group G, and Z is the remainder of X. Then we assume that the steric parameter υ_X is given by the equation

$$\upsilon_{X^1} = f_{M^1} + f_Z \tag{35}$$

Consider a second group X^2 such that $X^2 = M_0^1Z$. Thus, X^1 and X^2 have the same Z but different M^1, and

$$\upsilon_{X^2} = f_{M_0^1} + f_Z \tag{36}$$

Then

$$\upsilon_{X^1} - f_{M^1} = \upsilon_{X^2} - f_{M_0^1} \tag{37}$$

or

$$\upsilon_{X^1} = \upsilon_{X^2} + f_{M^1} - f_{M_0^1} \tag{38}$$

If we consider a set of X^1 groups for which M^1 is constant and a set of M_0^1 groups for which M^2 is constant,

$$\upsilon_{X^1} = \upsilon_{X^2} + c \tag{39}$$

where c is a constant. If we are interested in correlating rate constants for acid hydrolysis of esters WCO_2Z, where the acyl substituent W is constant and Z varies, we would use the equation

$$\log Kr_{OZ} = S\upsilon_{OZ} + h \tag{40}$$

From the equation

$$\upsilon_{OZ} = \upsilon_{CH_2Z} + c \tag{41}$$

and

$$\log kr_{OZ} = S\upsilon_{CH_2Z} + S_C + h \tag{42}$$

$$= S\upsilon_{CH_2} + h' \tag{43}$$

Correlation of log kr with Eq. 43 supplied the S value required to define υ_{OZ} values. Values of $\upsilon_{ef,OZ}$ may now be calculated from Eq. 34 if h can be determined. To evaluate h from Eq. 40, it is necessary to have one value of $\upsilon_{ef,OZ}$. Successful correlation of the rate constants for acid hydrolysis of WCO_2Z with Eq. 43 conformed the validity of Eq. 35. For any groups M^1Z^1 and M^1Z^2

$$\upsilon_{M^1Z^2} - \upsilon_{M^1Z^1} = f_{M^1} + f_{Z^2} - f_{M^1} - f_{Z^1} \tag{44}$$

$$= f_{Z^2} - f_{Z^1} \tag{45}$$

$$\upsilon_{M^1Z^2} = \upsilon_{M^1Z^1} + f_{Z^2} - f_{Z^1} \tag{46}$$

Applying Eq. 45 to CH_2H and CH_2Me gives

$$f_{Me} - f_H = \upsilon_{CH_2Me} - \upsilon_{CH_2H} = 0.56 - 0.52 = 0.04 \tag{47}$$

Then from Eq. 40, the known value of υ_{CH}, and the value for $f_{Me} - f_H$ we obtain $\upsilon_{OMe} = 0.36$. From this value and the value of S obtained above, h in Eq. 40 can be calculated, With S and h known, Eq. 34 can now be used to obtain υ_{ef} values for OZ groups.

Values of υ_e for SZ and NZ^1Z^2 groups are obtained in a manner analogous to that used for OZ groups [27,28]. The method results in a common scale for all groups making possible the inclusion of a wide range of substituent type in the same data set. The υ_{ef} values for OZ, SZ, and NZ^1Z^2 groups are also given in Table 4.

For groups of type 1 with $M^1 = CH_2$, M^2 = H, F, Cl, Br, I, OH, NH_2, Me, we may calculate a value of $\bar{\upsilon}_{ef}$ of 0.585 ± 0.0555. This is not significantly different from $\upsilon_{mn,Me} = 0.52$. The effect of M^2 is small. This result supports the MSI principle. The value of $\bar{\upsilon}_{ef}$ obtained for the groups $M^1 = CH_2$, $M^2M^3 ... = OMe$,

OEt, SMe, SEt, Et, Pr, CH_2Ph, CH_2CH_2Ph is 0.676 ± 0.0366. Obviously the combined effect of M^2, M^3 ... is significant. The steric effect of a group is dependent on contributions from the entire group, and not simply from M^1. This is further supported by the correlation of υ_{ef} values for 1, with $M^1 = CH_2$, O, S, NH; M^2 = H, F, Cl, Br, I, CH_2, O, S, NH; M^3 = Me, H; with the equation

$$\upsilon_{ef,\,M^1M^2M^3} = a_1\upsilon_{mn,\,M^1} + a_2\upsilon_{mn,\,M^2} + a_3\upsilon_{mn,\,M^3} + a_0 \qquad (48)$$

The results are highly significant $a_1 = 0.988 \pm 0.0862$; $a_2 = 0.149 \pm 0.0359$; $a_3 = 0.199 \pm 0.0370$; $a_0 = -0.00155 \pm 0.0428$; $F = 62.39$; $R = 0.9622$; $S_{est} = 0.0351$; $\psi = 0.307$; $100R^2 = 92.58$.

Obviously the greatest contribution to υ_e is made by υ_{mn} of M^1. Furthermore, the contribution of M^2 is proportional to $\upsilon_{mn,\,M^2}$.

Eq. 48 forms the basis of the separation of steric effects into increments due to each M^i in the group.

3.6 Steric Parameters of LCD Groups

The largest category of LCD groups is that of planar π bonded (pπ) groups. There are two main categories of pπ groups, 2 and 3. In 2, both the MZ^1 and MZ^2 bonds have a bond order greater than one, whereas in 3, only the MZ^1 bond has a bond order greater than one.

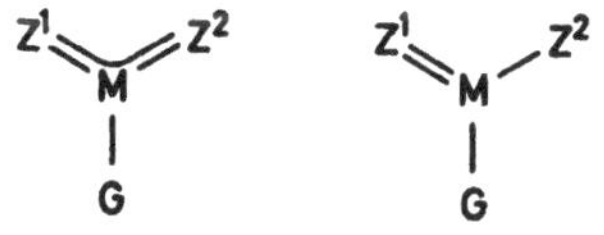

2 *3*

Examples of 2 are NO_2, Ph, and $CONH_2$; examples of 3 are Ac, CHO and Vi. The υ_{mn} value of either 2 or 3 is equal to its half thickness. Bondi has given r_V values for the half thickness of Ph(1.77), NO_2(1.55) and COZ(1.70). There are two possible values of υ_{mx} which must be considered when $Z^1 \neq Z_2$ (Fig. 11). The choice between these values can be made on the basis of the MSI principle, thus the smaller of the υ_{mx} values will be the one of interest.

If a planar π bonded group, Xpπ, is attached to a π bonded skeletal group, Gpπ, at a position adjacent to one bearing some other group, Y, then Xpπ is usually twisted through a dihedral angle, θ, due to steric repulsions between Xpπ and Y. The steric effect exerted by Xpπ depends on the angle θ. If the geometry of the systems is known, $r_{V,ef}$ and therefore υ_{ef}, can be calculated for Xpπ. In early work in the field no model of the geometry was available. It was therefore necessary to assume that the Xpπ group was in one of the two extreme conformations, the coplanar (θ = 0) or the perpendicular (θ = 90). For the former conformation $\upsilon_{ef} = \upsilon_{mx}$ while for the latter $\upsilon_{ef} = \upsilon_{mn}$. Not only is the steric effect of Xpπ a function of θ, its delocalized electrical effect also depends on θ. When θ = 0, π-delocalization is at a maximum whereas when θ = 90, it is zero. Thus, for the former case the delocalized electrical effect of Xpπ is equal to $\sigma_{D,Xp\pi}$, while in the latter it is equal to 0. It follows, then, that when $0 < \theta < 90°$, $\sigma_{DO} > \sigma_{D\theta} > 0$ where σ_{DO} is the value of σ_D observed when the dihedral angle is 0, $\sigma_{D\theta}$ is the value of σ_D when the dihedral angle is θ, and 0 the σ_D value when the dihedral angle is 90°.

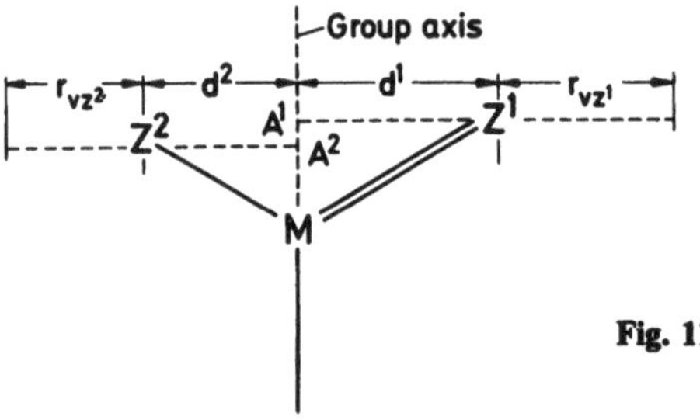

Fig. 11. A planar Π bonded group MZ^1Z^2

A model for the geometry of this type of system has now been developed [29]. Consider an ortho substituted benzene, 4, bearing an Xpπ group, MZ^1Z^2 and a nonplanar substituent, Xnp. The proposed geometry of 4 is shown in Fig. 12. We may define the group axis of a substituent as a line collinear with the Xpπ—G bond. Then, A is the intersection of the Xpπ group axis with the line segment Z^1Z^2; B is the intersection of the line through Xnp perpendicular to Z^1Z^2 with Z^1Z^2. Clearly,

$$\overline{BXnp} = r_{V,Xnp} + r_{V,mn,Xp\pi} = r + r' \qquad (49)$$

From Fig. 11

$$\overline{AX} = l_{CC} \cdot \cos 30° + l_{CXnp} \cdot \cos 53° \qquad (50)$$

where l_{CC} and l_{CX} are the bond lengths of the C—C and C—X bonds respectively. The length of any bond may be written as the sum of the covalent radii, r_C, of the bonded atoms. Then,

$$l_{CXnp} = r_{C,C} + r_{C,Xnp}; \qquad l_{CC} = 2r_{C,C} \qquad (51)$$

Thus,

$$\overline{AX} = (3r_{CC} + r_{CX}) \cos 30° \qquad (52)$$

It is well known that covalent radii are a linear function of van der Waals radii.

$$r_{VX} = r_{CX} + b \qquad (53)$$

Then, from Equations 52 and 53,

$$\overline{AX} = (3r_{CC} - b) \cos 30° + \cos 30° \, r_{VX} \qquad (54)$$
$$= a_{11}r + a_{10}$$

As

$$\sin\theta = \frac{\overline{BX}}{\overline{AX}} = \frac{r + r'}{a_{11}r + a_{10}} \tag{56}$$

Then

$$1/\sin\theta = a_{11} + \frac{(a_{10} - a_{11} \cdot r')}{r + r'} \tag{57}$$

As r′ is constant within a data set

$$1/\sin\theta = \frac{a_{30}}{r \cdot r'} + a_{11} = \frac{a_{30}}{\Sigma r} + a_{11} \tag{58}$$

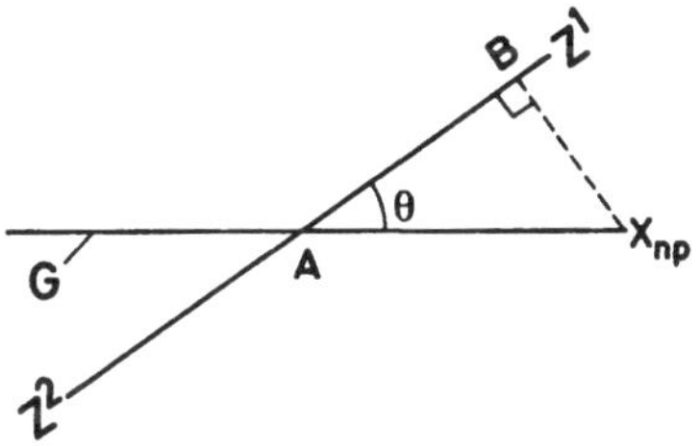

Fig. 12. End view of the geometry of an ortho-substituted benzene derivative. The X_{np} group is variable, the planar Π bonded MZ^1Z^2 group is constant

Values of θ measured by means of X-ray diffraction, dipole moment, electron diffraction, ultraviolet spectroscopy, and nmr spectroscopy for various sets of XpπGXnp with G = ortho phenylene or cis-vinylene were successfully correlated with Eq. 58, supporting the validity of the model.

We now consider the problem of the calculation of delocalized electrical effect substituent constants, σ_D, which correspond to a particular value of the dihedral angle θ, of the Xpπ group. From the geometry previously determined for this type of system (Fig. 13) we may define the effective Van der Waals radius of the Xpπ group MZ^1Z^2 as

$$r_{V,X_p} \equiv d' + r_{VZ^1} \tag{59}$$

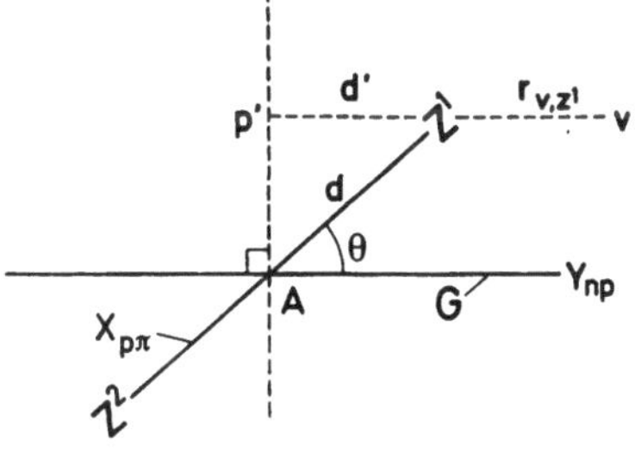

Fig. 13. Top view of the geometry of an ortho-substituted benzene derivative

when $r_{VZ^1} < r_{VZ^2}$. The line PP″ through the group axis of Xpπ is perpendicular to the plane of Gpπ. The group axis of Xpπ is collinear with the Xpπ—Gpπ bond. The line P′V is the perpendicular through Z^1 to PP″. Then, $d = P^1Z^1$.

The steric parameter υ was defined as

$$\upsilon_X \equiv r_{VX} - r_{VH} = r_{VX} + 1.20 \tag{17}$$

Then

$$\upsilon_{Xp\pi} = d' + r_{V^1} - 1.20 \tag{60}$$

Clearly,

$$d' = d \cos \theta \tag{61}$$

where d is the distance between the point of intersection of the perpendicular through Z^1 to the group axis and Z^1 (Fig. 14). From Eq. 60 and 61

$$\upsilon_{Xp\pi} = d \cos \theta + r_{VZ^1} - 1.20 \tag{62}$$

or

$$\cos \theta = (\upsilon_{Xp\pi} - r_{VZ^1} + 1.20)/d \tag{63}$$

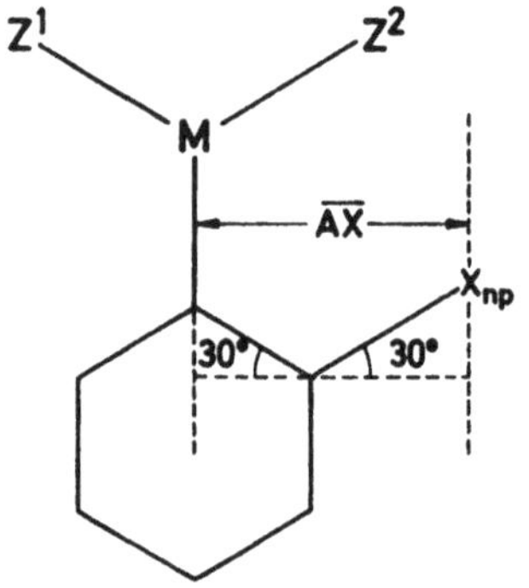

Fig. 14. The geometry of an ortho substituted benzene derivative with constant Y_{np} and variable MZ^1Z^2

Eq. 62 and 63 show the dependence of the steric parameter for an Xpπ group upon the dihedral angle, θ. The σ_D constant of an Xpπ group must also depend upon the dihedral angle since the extent of delocalization is a function of θ.

Most workers who have estimated values of θ have made use of an equation of the type

$$Q = Q_0 \cos^2 \theta \tag{64}$$

or one derived from it. From Eq. 64 with $Q = \sigma_D$,

$$(\sigma_{D\theta}/\sigma_{DO})^{1/2} = \cos \theta \tag{65}$$

Substitution of Eq. 63 in Eq. 65 gives

$$(\sigma_{D\theta}/\sigma_{DO})^{1/2} = (\upsilon_{Xp\pi} - r_{VZ} + 1.20)/d \qquad (66)$$

As $\sigma_{D\theta}$ must lie between O and σ_{DO}, both it and σ_{DO} must have the same sign. Then,

$$|\sigma_{D\theta}/\sigma_{DO}|^{1/2} = (\upsilon_{Xp\pi} - r_{VZ} + 1.20)/d \qquad (67)$$

Rearranging 67

$$|\sigma_{D\theta}|^{1/2} = |\sigma_{DO}|^{1/2}\,(\upsilon_{Xp\pi} - r_{VZ} + 1.20)/d \qquad (68)$$

As $|\sigma_{DO}|^{1/2}$, r_{VZ}, and d are constant, Eq. 68 has the form

$$|\sigma_{D\theta}|^{1/2} = a_1\,{}_{Xp\pi} + a_0 \qquad (69)$$

where

$$a_1 = |\sigma_{DO}|^{1/2}/d\,, \qquad a_0 = (1.20 - r_{VZ})\,|\sigma_{DO}|^{1/2}/d \qquad (70)$$

In order to include an $Xp\pi$ group in a data set in which the substituents exhibit an steric effect it is necessary to know θ so that the appropriate values of the delocalized electrical effect parameter, σ_D and the steric parameter υ can be used. Values of θ are usually unavailable. A method has been developed to circumvent this difficulty for chemical reactivities. The procedure is: 1) A basis set, which contains only those members of the data set that show minimal dependence of their steric effect on conformation, is correlated with the LDS equation,

$$Q_X = L\sigma_{IX} + D\sigma_{DX} + S\upsilon_X + h \qquad (71)$$

In Eq. 71, Q is the quantity to be correlated, and σ_I is a parameter which represents the localized electrical effect.
2) The data point for an $Xp\pi$ group is now added to the basis set. Separate correlations are carried out with Eq. 71 using values for $Xp\pi$ increasing in convenient increments from υ_{mn} to υ_{mx} (Fig. 9) and the corresponding σ_D values calculated from Eq. 69. The proper values of $\upsilon_{Xp\pi}$ and $\sigma_{D,\,Xp\pi}$ are those which result in the best correlation of the data set with Eq. 71. The best correlation is that which results in maximum values of statistics such as the F test or $100R^2$ (where R is the multiple correlation coefficient); or in minimum values of such statistics as the standard error of the estimate, S_{est}; or the ψ^* statistic (defined as the ratio of S_{est} to the root mean square of the data). To be meaningful, it is also necessary that the best correlation give values of the regression coefficients L, D, and S which are not significantly different from those obtained for the correlation of the basis set in Step 1.

Results of correlations of chemical reactivities and physical properties using this method have shown that frequently NO_2 and COX have υ_{ef} values which lie between

υ_{mn} and υ_{mx}. Generally, Ph takes the value υ_{mx}. These results apply to ortho substituted benzenes. When no group other than H is vicinal to a substituent NO_2 and COX are likely to be coplanar with any π bonded skeletal group to which they are bonded. The phenyl and substituted phenyl groups generally have υ_{ef} values between υ_{mn} and υ_{mx} in this case.

When an Xpπ group is bonded to an sp^3 hybridized carbon atom there is no delocalized electrical effect contribution and only the steric effect of the group need be considered.

3.7 The Degree of Dependence of υ_{ef} Values on Electrical Effects

We have noted above that if steric effects are to be successfully separated from other substituent effects, it is vital that steric parameters show a minimal dependence on electrical effect parameters or transport parameters. Unger and Hansch [30] have reported that E_S values of CH_2X groups are significantly dependent on electrical effects. To resolve this question for the υ_{ef} values we have carried out correlations of the $\upsilon(CH_2)_nZ$ values with the equation

$$\upsilon_e(CH_2)_nZ = a_{11}\sigma_I + a_{10} \tag{72}$$

using σ_I values taken from the compilation of Charton [33]. There is no need to include a term in σ_D as $\sigma_{D,(CH_2)_nZ}$ is a linear function of $\sigma_{I,(CH_2)_nZ}$. This follows from the relationship

$$\sigma_{I,CH_2Z} = b_{11}\sigma_{IZ} + b_{10} \tag{73}$$

and

$$\sigma_{D,CH_2Z} = b_{21}\sigma_{IZ} + b_{20} \tag{74}$$

When n = 1 and Z = H, F, Cl, Br, I, Me, Et, iPr, tBu, OH, OMe, OEt, OPr, OiPr, CH_2Vi, SMe, SEt, Ph, Vi, CH_2Ph, CH_2CH_2Ph, CO_2Me, NO_2, there is no significant correlation between υ_{ef} and σ_I (set 1, Table 6). When n = 3 and Z = H, Me, Et, OMe, OEt, OPr, Ph, Vi and $PhCH_2$, there is again no significant correlation (Set. 3, Table 6). Correlation of υ_{ef,CHZ^1Z^2} with the equation

$$\upsilon_{ef,X} = a_{21}\Sigma\sigma_{IZ} + a_{20} \tag{75}$$

also gave no significant correlation (set 4, Table 6), where Z^1, Z^2 = Me, Me; Me, Et; Et, Et; Ph, Me; Ph, Et; Ph, Ph; Ph, OH; Me, OH; Me, OEt; H, H; Me, EtS; EtS, EtS; Me, Br; Me, Cl; Ph, Cl; F, F; Cl, Cl; Br, Br; I, I. The same result was obtained on correlation of $\upsilon_{ef,CZ^1Z^2Z^3}$ with Eq. 75 (set 5, Table 6), with $Z^1Z^2Z^3$ = Me, Me, Me; Me, Me, Et; Et, Et, Et; tBu, Me, Me; Ph, Ph, Me; Ph, Ph, Et; Ph, Ph, Ph; Me, Me, Br; Me, Br, Br; F, F, F; Cl, Cl, Cl; Br, Br, Br; I, I, I; Me, Me, OH; H, H, H. Generally then υ_{ef} values for CZ_nH_{3-n} are independent of electrical effects. There is one notorious exception to this conclusion however. For $\upsilon_{ef,(CH_2)_2Z}$ a highly significant correlation with Eq. 72 is obtained, where Z is H, Me, Et, iPr, tBu, Cl,

Br, I, OH, OMe, OEt, OPr, OiPr, SMe, SEt, Ph, $PhCH_2$, $PhCH_2CH_2$, Vi, $ViCH_2$, Ac, NO_2 (set 2, Table 6). Best results are obtained on the exclusion of the value for X = H (Set 2A). The question arises as to the cause of this unusual behavior.

The conformation of 1-substituted-3,3-dimethylbutanes seems to depend in part on σ_I. Thus, ΔE_X for the equilibrium between gauche, 4, and trans, 5, conformers, determined by Whitesides, Sevenair, and Goetz,[31] have been correlated with the equation[32]

$$\Delta E_X = L\sigma_{IX} + S\upsilon_X + A\alpha_X + h \tag{76}$$

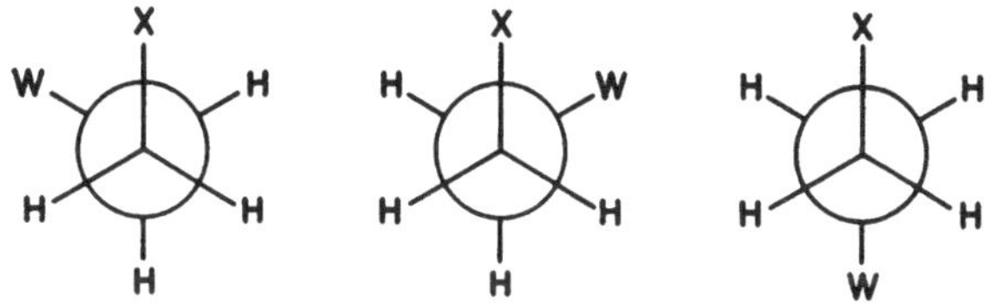

4a 4b 5

4, 5; W = tBu 5, 7; W = G

If the tBu group in 4 and 5 is replaced by the skeletal group, G, we have a system which models that from which the $\upsilon_{CH_2CH_2X}$ constants have been derived. As the steric effect of a CH_2CH_2X group is conformationally dependent and the conformation is a function at least in part of σ_I, the dependence of υ_{ef,CH_2CH_2X} on σ_I is reasonable.

Table 6. Results of Correlations with Equations 72, 75, and 83

Set	a_{n_1}	a_{n_0}	r^a	F^b	S_{est}^c	$S_{a_{n_{11}}}^c$	$S_{a_{n_{10}}}^c$	ψ^{*e}	n^f	$100r^{2\,d}$
1	−0.171	0.745	0.2016[m]	0.890[m]	0.171	0.182[n]	0.0535	1.025	23	4.06
2	0.586	0.674	0.9069	87.97	0.0463	0.0625	0.0153	0.443	21	82.24
2A	0.543	0.688	0.9235	104.4	0.0381	0.0532	0.0133	0.404	20	85.29
3	0.0269	0.693	0.1633[m]	0.192[m]	0.0230	0.0614[o]	0.0107	1.119	9	2.67
4	−0.124	0.977	0.1524[m]	0.404[m]	0.290	0.195[p]	0.100	1.045	19	2.32
5	−0.300	1.85	0.2495[m]	0.863[m]	0.737	0.323[n]	0.256	1.040	15	6.22
6	2.59	0.845	0.9825	249.8	0.0547	0.164	0.0368	0.206	11	96.52
7	3.79	0.263	0.9219	45.30	0.0680	0.564	0.0425	0.433	10	84.99

For footnotes not given below, see Table 3
m. 90.0% CL. n. 60.0% CL. o. 30.0% CL. p. 40.0% CL.

3.8 The Degree of Dependence of the Steric Parameter on Transport Parameters

The most frequently used transport parameter is log P_{XGY}, the logarithm of the partition coefficient of the substrate XGY between water and some solvent which is presumed to model a biomembrane or a biopolymer. A very large number of log P values are available for the system water: 1-octanol. An alternative to the use of log P is the use of the parameter π defined by [34] the expression

$$\pi_X = \log P_X - \log P_H \tag{76}$$

Another approach lies in the use of the R_M parameter, derived from chromatography. R_M is defined by [35]

$$R_M \equiv \log [(1/R_f) - 1] \tag{77}$$

It had previously been shown that [36]

$$P = K[(1/R_f) - 1] \tag{78}$$

where K is a constant characteristic of the chromatographic system. Thus,

$$\log P = \log K + R_M \tag{79}$$

As log P values in one solvent pair are related to those in other solvent pairs [36] by the equation

$$\log P = a_1 \log P^\circ + a_0 \tag{80}$$

where log P° values are for the reference solvent pair. Thus, R_M, π and log P are all inter-related. Charton and Charton [38] have shown that log P, and by inference, π values for amino acids are a function of the difference in intermolecular forces between amino acid and water and between amino acid and other solvents. The correlation equation was of the form

$$\log P_X = A\alpha_X + L\sigma_{IX} + S\upsilon_X + H_1 n_{HX} + H_2 n_{nX} + Ii_X + B_0 \tag{81}$$

where X is the amino acid side chain, α is the polarizability parameter defined by

$$\alpha_X = (M_{RX} - M_{RH})/100 \tag{82}$$

σ_I is the localized effect parameter, υ the effective steric parameter, n_H is equal to the number of NH and/or OH bonds in X, n_n the number of full nonbonding valence shell orbitals on N or O atoms in X, and i is 1 for an ionic side chain, O for a nonionic side chain. The α and σ_I parameters account for dipole-dipole (σ_I), dipole-induced dipole(σ_I,α) and induced dipole-induced dipole(α) interactions. The n_H and n_n parameters account for hydrogen bonding. The steric parameter υ represents the steric effect of the side chain on the solvation of the α-amino and carboxyl groups of the amino acid. It seems not unreasonable that R_M and log P values for many other types of compound are also represented by Eq. 81 or relationships derived from it. Then from Eq. 81 it seems quite likely that a significant degree of correlation between steric parameters and transport parameters will frequently be observed. In further support of this conclusion, υ_{mn} for MZ_3 groups including CF_3, CBr_3, CCl_3,

CMe_3, SiF_3, $SiCl_3$, $SiBr_3$, $SiMe_3$, GeF_3, $GeCl_3$, and $GeBr_3$ are well correlated by the equation (X = M_3)

$$\upsilon_X = a_{31}\alpha_X + a_{30} \tag{83}$$

Results of the correlation are given in Table 6 (set 6). Values of υ_{mn} for NCD groups and $M(lp)_nH_{3-n}$ groups are also correlated successfully with Eq. 83 with X = F, Br, Cl, I, C≡CH, CN, OH, NH_2, SH, PH_2 groups. (Set 7, Table 6).

The problem of significant correlation of the transport with the steric parameter can be minimized by a proper choice of substituents. Thus, $nAk[-CH_2)_nH]$ groups with $n \geqq 3$ have a constant steric effect ($\upsilon_{ef} = 0.68$) and localized electrical effect ($\bar{\sigma}_I = 0.01$), but their polarizability α, is a linear function of n. Alkyl groups with the same number of carbon atoms have the same values of α and σ_I but different υ_{ef} values. Inclusion of these groups in the data set will decrease the collinearity between steric and transport parameters. Alternatively, they may be treated as separate subsets to establish clearly dependence on steric effects or on transport. Thus, successful correlation of the biological activities of a set of nAk groups with a transport parameter establishes a dependence on the transport parameter as the electrical and steric effects are constant. Successful correlation of the biological activities of a set of Ak groups for which the number of carbon atoms is constant will establish a dependence on steric effects, as electrical and transport effects are constant.

4 Summary

We have described the Van der Waals radii, r_V, the definition of υ from r_V for symmetric groups, and the conformational dependance of the steric effects of non-symmetric groups. The experimental determination of υ values for nonsymmetric groups from rate constants for reactions whose structural dependance is limited to steric effects was discussed. A method was presented for handling the steric and delocalized electrical (resonance) effects of planar π bonded groups. The magnitude of the gear effect for MZ_3 groups was shown to be generally small. This work provides a basis for the treatment by correlation analysis of steric effects upon chemical reactivities, physical properties, and biological activities.

5 Appendix 1

5.1 Derivation of Group Van der Waals Radii

Derivation of r_{mx}, r_{ax}, l_{MG}, for a group MZ_3 with M hybridized sp^3.

1. Draw the tangent line through B normal to group axis $\overline{OA}$ where O is the center of the M atom.
2. Draw line $\overline{CD}$ through E parallel to $\overline{AB}$

3. $\overline{OE}$ is l_{MZ} (MZ bond length)
4. θ is $\measuredangle$ EOG (ZMG bond angle).
5. $\eta = 180° - \theta = \measuredangle$ EOD.
6. $\overline{CD}$ is r_{mx}.
7. $\overline{CD} = \overline{CE} + \overline{ED}$.
8. $\overline{CE} = r_{V,Z}$.
9. $\overline{ED} = l_M \sin\eta$.
10. $r_{mx} = r_{VZ} + l_{MZ} \sin\eta$.
11. $r_{mx} = \overline{AD} + \overline{OD}$.
12. Construct $\overline{EB}$ parallel to $\overline{AD}$.
13. $\overline{EB} = \overline{AD} = r_{V,Z}$.
14. $\overline{OD} = l_{MZ} \cos\eta$.
15. $r_{ax} = r_{VmZ} + l_{MZ} \cos\eta$.
16. Extend $\overline{OA}$ to G (the center of atom G to which MZ_3 is attached).
17. $l'_{MG} = \overline{OG} + \overline{OD} = l_{MG} + l_{MZ} \cos\eta$, (where l_{MG} is the MG bond length).

The construction is shown in Fig. 15.

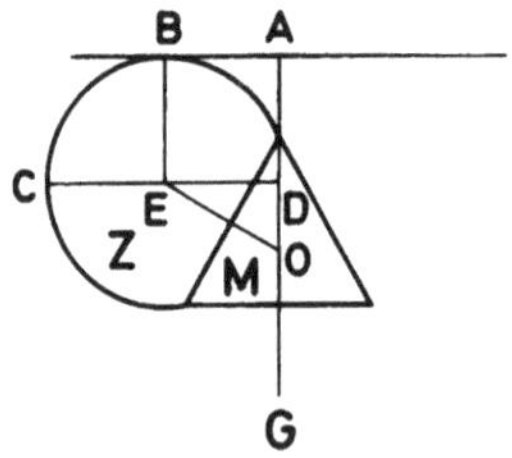

Fig. 15. Construction for the derivation of $r_{V,mx}$; $r_{V,ax}$ and l_{MG} for MZ_3 groups

5.2 Derivation of r_{mn} for MZ_3 where M is hybridized sp^3

1. Construct the line $\overline{HI}$ tangent to atoms Z^1 and Z^2.
2. Construct the normal to $\overline{IH}$ through D which lies on the group axis ($\overline{AG}$ in Fig. 14).
3. Construct $\overline{EH}$ normal to $\overline{HI}$ through E, the center of Z^1.
4. Construct $\overline{EK}$ normal to $\overline{DJ}$ through E.
5. $r_{V,mn} = \overline{JD} = \overline{JK} + \overline{KD}$.
6. $\overline{JK} = \overline{EH} = r_{V,Z}$.
7. Construct $\overline{LD}$.
8. $\measuredangle$ KDL = 120°.
9. $\measuredangle$ KDE = 60°.
10. $\overline{KD} = \overline{ED} \cos 60° = \overline{ED}/2$.
11. $\overline{ED} = l_{MZ} \sin\eta$.
12. $r_{V,mn} = r_{V,Z} + (l_{MZ} \sin\eta)/2$.

The construction is shown in Fig. 16.

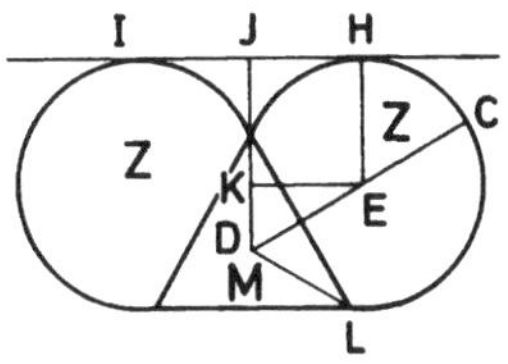

Fig. 16. Construction for the derivation of $r_{V,mn}$ for MZ_3 groups

6 References

1. Taft, R. W.: The Separation of Polar, Steric, and Resonance Effects in Reactivity, in: Steric Effects in Organic Chemistry (ed. Newman, M. S.) Wiley, New York 1956
2. Pauling, L.: The Nature of the Chemical Bond, Cornell University Press, Ithaca 1960³
3. Bondi, A.: J. Phys. Chem. *68*, 441 (1964)
4. Bondi, A.: J. Phys. Chem. *80*, 3006 (1966)
5. Charton, M.: J. Am. Chem. Soc. *91*, 615 (1969)
6. Charton, M.: Prog. Phys. Org. Chem. *10*, 81 (1973)
7. Allinger, N. L.: Adv. Phys. Org. Chem. *13*, 1 (1976)
8. Bartell, L. S.: J. Chem. Phys. *32*, 827 (1960)
9. Glidewell, C.: Inorg. Chim. Acta *12*, 219 (1975)
10. Glidewell, C.: Inorg. Chim. Acta *20*, 113 (1976)
11. Charton, M.: J. Am. Chem. Soc. *101*, 7356 (1979)
12. Charton, M., Charton, B. I.: J. Org. Chem. *36*, 260 (1971)
13. Charton, M.: J. Org. Chem. *36*, 266 (1971)
14. Charton, M.: J. Org. Chem. *36*, 882 (1971)
15. Hawkins, B. L., Bremser, W., Borcic, S., Roberts, J. D.: J. Am. Chem. Soc. *93*, 4472 (1971)
16. Roussel, C., Chanon, M., Metzger, J.: Tetrahedron Letts. 1861 (1971)
17. Roussel, C., Chanon, M., Metzger, J.: FEBS Letts. *29*, 253 (1073)
18. Liden, A., Roussel, C., Chanon, M., Metzger, J.: Tetrahedron Letts. 3629 (1974)
19. Carter, R. E., Drakenberg, T., Roussel, C.: J. Chem. Soc. Perkin Trans. II, 1690 (1975)
20. Roussel, C., Liden, A., Chanon, M., Metzger, J., Sandstrom, J.: J. Am. Chem. Soc. *98*, 2847 (1976)
21. Liden, A., Roussel, C., Liljefors, T., Chanon, M., Carter, R. E., Metzger, J., Sandstrom, J.: J. Am. Chem. Soc. *98*, 1976 (1976)
22. Nilsson, B., Martinson, P., Olsson, K., Carter, R. E., J. Am. Chem. Soc. *96*, 3190 (1974)
23. Oki, M.: Angew. Chem. Intl. Edn. *16*, 87 (1976)
24. Charton, M.: J. Am. Chem. Soc. *98*, 1552 (1975)
25. Charton, M.: J. Org. Chem. *41*, 2217 (1976)
26. Charton, M.: J. Org. Chem. *42*, 3531 (1977)
27. Charton, M.: J. Org. Chem. *42*, 3535 (1977)
28. Charton, M., Charton, B. I.: J. Org. Chem. *43*, 1161 (1978)
29. Charton, M.: J. Org. Chem. *48*, 1011, 1016 (1983)
30. Unger, S., Hansch, C.: Prog. Phys. Org. Chem. *12*, 91 (1976)
31. Whitesides, G. M., Sevenair, J. P., Goetz, R. W.: J. Am. Chem. Soc. *89*, 1135 (1967)
32. Charton, M.: The Directing and Activating Effects of Triply Bonded Groups, in: The Chemistry of Functional Groups, Supplement C. (ed. Patai, S., Rappaport, Z.) Wiley, New York, p. 269
33. Charton, M.: Prog. Phys. Org. Chem. *13*, 119 (1981)
34. Hansch, C., Muir, R. M., Fujita, T., Maloney, P. P., Geiger, F., Streich, M.: J. Am. Chem. Soc. *84*, 2817 (1963)
35. Bate-Smith, E. C., Westall, R. G.: Biochem. Biophys. Acta *4*, 427 (1950)
36. Martin, A. J. P.: Biochem. Soc. Symp. Camb. *3*, 4 (1949)
37. Collander, R.: Acta Chem. Scand. *6*, 774 (1954)
38. Charton, M., Charton, B. I.: J. Theoret. Biol. 99, 629 (1982)

Molecular Shape Descriptors

Ioan Motoc

Institut für Strahlenchemie, 4330 Mühlheim/Ruhr 1, FRG

Table of Contents

This section reviews the molecular shape descriptors developed by Amoore, Allinger, Simon et al. and Testa and Purcell. The illustrative examples discussed refer to the odour similarity and cardiotoxic aglycones. One has stressed the methods based on the reference structure because, correctly formulated, these methods seem to offer promising perspectives to model the steric effects in biological systems. Finally, a short discussion of possible connections between steric and other substituent constants (relevant in the context of multicollinearity in QSAR) is included.

1 Comparison of Molecular Shape Based on a Reference Structure

The methods discussed here (see also [1,2]), namely those developed by Amoore, Allinger, and Simon et al., are based on the use of a reference structure to compare the molecular shapes. They differ through the option adopted to perform the geometrical congruence of the reference structure *vs.* compared structure, and through the manner in which one expresses numerically the shape dissimilarity. The resultand dissimilarity coefficients cannot be directly compared with other steric parameters because $\bar{\Delta}$ (Amoore) and A (Allinger) are calculated for molecules while most steric constants available refer to substituents, and MSD, MCD and MTD (Simon et al.) depend on the series of biomolecules being considered. In principle, all these five parameters can be used for series of widely differing structures.

1.1 Amoore's Approach (the $\bar{\Delta}$ Parameter)

This method has been developed for the research on the stereochemical factors influencing the sence of smell [3-5,7] (i.e., the quantitative formulation for Ruzicka's stereochemical theory [6] of olfaction). The comparison of the molecular shape is made by means of molecular model silhouettes within a procedure which consists of:

i) "front", "top", and "right"-silhouettes (i.e., projections of space-filling models on three mutually orthogonal planes) of the two molecules (the reference and the compared structure, respectively) are obtained.

ii) pairs of silhouettes of the same type are superimposed respecting superposition of weight centres and collinearity of main axis in both cases.

iii) on the silhouettes thus superimposed thirty-six radii are traced with 10° angular spacing, from the weight center toward periphery. Absolute differences (in Å) between corresponding radii in silhouette pairs are added to yield the Δ value.

iv) the average of the three Δ values for the three types of silhouettes gives the Amoore's $\bar{\Delta}$ parameter. $\bar{\Delta}$ may be normalized as:

$$\bar{\Delta}_N = 1/(1 + \bar{\Delta})$$

On the basis of the method described above, the PAPA machine (i.e., *P*robabilistic *A*utomatic *P*attern *A*nalyser — a combination of photocamera/interface/computer) has been developed by Amoore, Palmieri and Wanke [4,8], and applied in a study of ant alarm pheromone activity. The 36 radii used for hand calculation of $\bar{\Delta}$ are substituted within PAPA by a reproducible collection of 4096 randomly selected curves.

Recently, Motoc et al. have studied [9] the correlational ability of Amoore's parameter, the stability and predictive value of the resulted equations (the odour similarity, Y, and the corresponding $\bar{\Delta}_N$ values were taken from ref. [10]). The model:

$$Y = a + b\,\bar{\Delta}_N \tag{1}$$

has been calibrated using 30 odorants, while 10 different odorants were used for cross-validation. Recomputing the equation (1) for the whole series of 40 odorants, one may estimate the model stability. The results are collected in Tabele 1.

Table 1. Correlation ability of $\bar{\Delta}_N$ parameter. Stability and predictive value of equation (1)

Y_i	a	b	r	s	F	n
Y_1	−5.486	12.834	0.62	1.25	8.33	30
Y_1	−0.908	5.388	0.64	1.45	3.67	40
Y_1	−1.635	1.626	0.80	0.99	6.26	10
Y_2	−1.703	6.254	0.61	0.48	7.88	30
Y_2	−1.760	6.383	0.67	0.48	13.79	40
Y_2	0.010	1.023	0.72	0.48	3.67	10
Y_3	−0.240	2.848	0.34	0.41	1.75	30
Y_3	−0.456	3.315	0.36	0.41	2.75	40
Y_3	−1.725	2.679	0.50	0.41	1.14	10
Y_4	−3.955	11.568	0.67	0.79	11.28	30
Y_4	−4.556	12.516	0.70	0.75	16.46	40
Y_4	−0.999	1.327	0.63	0.49	2.32	10
Y_5	−1.692	6.282	0.57	0.53	6.21	30
Y_5	−1.742	6.222	0.53	0.53	7.39	40
Y_5	0.090	0.749	0.41	0.46	0.70	10

(Y_i stand for the odour similarity: i = 1-ethereal, 2-camphoraceous, 3-musky, 4-floral, 5-minty)

The correlations which emerge are modest. They may be justified by the imprecision of experimental data (these data may be poor because odour similarity is obtained as average from a jury of 29 persons), or/and by the "ad-hoc" criterion used to perform the geometrical congruence.

1.2 Allinger's Approach (the A Parameter)

To compare the shape of two molecules, Allinger proposed [11] the following procedure: each molecule is placed in its own separate cube which measures 10 Å on a side. The molecules are described by the cartesian coordinates and van der Waals radii of their atoms. The 10 Å cube is divided into one million elementary cubes (each elementary cube measures 0.1 Å on a side). Then, the reference molecule and the compared molecule are superimposed by the criterion that one superimposes the moments of inertia of the two molecules. A comparison of elementary cube by elementary cube is made between the two 10 Å cubes, and the following occurences are considered significant: i) both 10 Å cubes are occupied at a given elementary cube (a match), or ii) one cube is occupied at a given elementary cube and the second one is not (a mismatch). Counting the total number of matches (denoted here by M) and the total number of mismatches (denoted by N), the goodness of the shape fit is expressed as:

$$A = M/(M + N) \quad (2)$$

The correlations obtained with the intensity of the odour of fatty acids (the reference structure is the isovaleric acid) were [11]: "not nearly as good as one would hope for".

The present author suggested [12] that poor correlations with the A parameter are due to its structure. It is easily seen that the above procedure simulates the Monte Carlo technique [13]. Accordingly, we proved [13] that A reads as:

$$A = OV/(OV + MCD) \quad (3)$$

where OV $\sim 10^{-3}$ M is the overlapping volume [14] of the two molecules, and MCD $\sim 10^{-3}$ N is the nonoverlapping volume [15] of the two molecules. Having an intricate structure, the A parameter is of questionable utility for QSAR.

The "4π-comparator" method [23,24] represents a qualitative version of the above two methods. One used CPK molecular models and the computer generates the six projections corresponding to the faces of the cube containing the respective pair of molecules. These projections reveal shape and volume differences among the molecules considered.

1.3 Simon et al.'s Approach (the MSD, MCD and MTD Parameters)

The Minimal Steric Difference method (bbreviated as MSD) offer [16a] a very simple way to express by a single number the molecular shape dissimilarity. As reference structure (which is termed here "standard", denoted by S_0) one selects generally the molecule of highest activity from the data base under study. The shape of the standard is considered (approximately) complementary to the receptor cavity. The geometrical congruences of standard *vs.* compared molecule are performed [16e] seeking the maximal superposition of the two molecules (small differences in bond lengths and bond angles are neglected, and the various conformations of flexible molecules are considered to be degenerate). The MSD parameter is defined as the number of non-superposable, non-hydrogen atoms of the two compared molecules.

For the structures depicted in Fig. 1, the MSD's are: MSD (S_1, S_2) = 1, MSD (S_1, S_3) = 1, MSD (S_4, S_5) = 2, MSD (S_3, S_2) = 0 etc.

When tested in correlations, the MSD parameter offered rather modest results (one should consult Ref. [16a] and the literature cited there in). Quite possibly, these results can be justified by the fact that a single number steric parameter does not

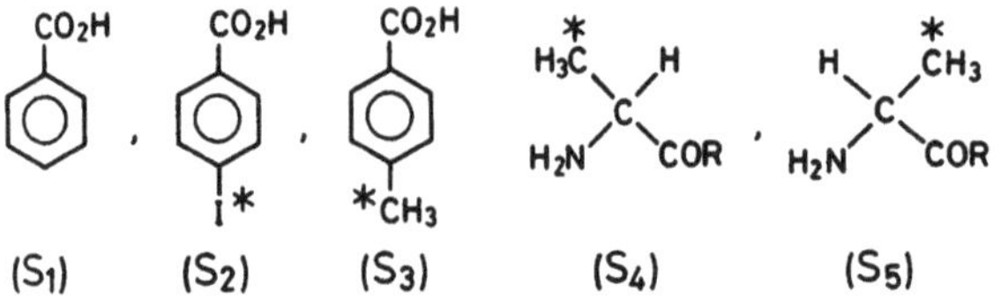

Fig. 1. Calculation of MSD steric parameter: illustrative example

quantify properly the steric interaction pattern in the context of the key-in-lock theory.

The MCD parameter [16b] "translates" the topological MSD parameter into the metric context: the molecules are described by cartesian coordinates and van der Waals radii of their atoms. The way one achives the superposition is implicitly specified by the atomic coordinates because all considered molecules are represented in the same cartesian coordinate system. In order to compute the total volume of the unsuperposable van der Waals envelopes, i.e., the MCD, one uses the Monte Carlo technique [13] as follows [15]: let a be the volume of the parallelipiped which contains the superimposed van der Waals envelopes of the two molecules. One generates a sequence of N_T random points uniformly repartized within this parallelipiped. Then, one selects from among the N_T points the N_I random points which are within the van der Waals envelope of one of the molecule but not within the other one. The MCD value is given by:

$$\mathrm{MCD} = a\, N_I/N_T$$

The correlations in terms of MCD were not improved relative to the MSD ones.

The Minimal Topological Difference (abbreviated as MTD) method denotes [16c] a "hypermolecule optimization" procedure based on the MSD parameter. Actually, MTD is the MSD parameter calculated [17–19] against a "hypermolecule" as standard.

The MTD procedure consists in the following steps:

i) One obtains the hypermolecule, $\hat{H}$, by approximative atom per atom superposition of the all molecules of the data base over the standard molecule, seeking maximal superposition. Occasionally, this rule is supplemented with other criteria such as: the reactive groups, or the hydrogen-bonding groups etc., of each molecule should be superimposed.

ii) $\hat{H}$ is used as topological framework for describing each molecule i (involved in the building of the hypermolecule) by the vector $\mathbf{x}_i = [x_{ij}]$. The entry x_{ij} is taken to be 1 if the vertex j of $\hat{H}$ is occupied by a non-hydrogen atom of the molecule i, and $x_{ij} = 0$ if it is not occupied.

iii) The initial receptor map, S^0, is derived from $\hat{H}$ as follows: the vertices j of $\hat{H}$ corresponding to the compound of the highest activity in the data base (i.e., the standard S_0) are assigned to the receptor cavity. They are characterized by the parameter $\varepsilon_j = -1$. Vertices j corresponding to the receptor walls are characterized by $\varepsilon_j = +1$, and those corresponding to the exterior of the receptor (i.e., the irrelevant space) by $\varepsilon_j = 0$. The $\varepsilon_j = +1$ or 0 assignements are made largely subjectively.

iv) The MTD value for the molecule i, MTD_i, is given by:

$$\mathrm{MTD}_i = s + \sum_{j=1}^{m} \varepsilon_j x_{ij} \tag{4}$$

where m and s stand for the number of the hypermolecule vertices, and the number of cavity vertices, respectively. The MTD_i signifies the number of relevant, non-superposable atoms of the molecule i and hypermolecule $\hat{H}$.

v) The regressional equation:

$$\hat{Y}_i = a + b\,\mathrm{MTD}_i \tag{5}$$

is calculated using the least squares method and the initial map S^0.

One changes the assignment ε_t^0 corresponding to the vertex t of S^0 (i.e., $\varepsilon_t^0 = -1, 0,$ or $+1$) by ε_t^* (i.e., $\varepsilon_t^* = +1$, or 0; -1, or 0; -1, or $+1$, respectively) calculating the amount $\Delta Y(\varepsilon_t)$ as:

$$\Delta Y(\varepsilon_t) = 2b(\varepsilon_t^* - \varepsilon_t^0) \sum_{i=1}^{N_{occ}} (Y_i - \hat{Y}_i) + \frac{(\varepsilon_t^* - \varepsilon_t^0)^2\, b^2 N_{t,emp} N_{t,occ}}{n} \tag{6}$$

Y and $\hat{Y}$ are the observed and the estimated (eq. 5) biological response, and $N_{t,emp}$ and $N_{t,occ}$ stand for the number of molecules in which the vertex t is empty and occupied, respectively. n is the number of molecules under consideration, and b is the regession coefficient with MTD in eq. 5. The substitution of ε_t^* for ε_t^0 which yiedds the most negative $\Delta Y(\varepsilon_t)$ value is effectively performed. The resulted map is considered as initial map and the steps (iv) and (v) are resumed.

vi) The procedure is stopped when all $\Delta Y(\varepsilon_t)$'s are non-negative on new monosubstitutions. The resulted map is termed optimal and denoted by S*. The $\varepsilon_j = -1$ vertices within S* define the optimal standard (the listing of the program implementing the above algorithm is published [16d]).

Many QSAR applications of the MSD and MTD parameters have been published: Simon and collaborators (one should consult the review articles [16a, e]), Duvaz et al. [25a], Carles and Goursot [26], Verloop [25b].

The present version of the MTD approach has serious deffects which cast much doubt concerning the realiability of these calculations. The most important points are systematized below:

1) It is easy to observe that eqn. (5) attributes the whole variance of the observed biological response to the MTD parameter and, because the MTD's change during the optimization procedure, the "optimal" MTD values will code globally the steric and non-steric factors which may control the biological response. One should be aware that, in order to use MTD as a steric parameter, it is mandatory to sort out the non-steric factors within biological response, and subsequently to perform MTD calculations.

2) The MTD algorithm is inconsistent, namely the conclusion contradicts the premise, i.e., the initial standard is a chemical compound while the optimal standard is a pattern of vertices which cannot be regarded as a molecule. This situation occurs because the MTD algorithm does not use all necessary information which characterizes the molecular shape: the number and the identity of the atoms *and* the molecular connectivities (i.e., the chemical bonds existing among atoms). In actual practice this method produces artifacts [20], e.g. high correlations irrespective of the biological input. The recent suggestion [22] to consider as initial standard the vertices which are present prodominantly in active molecules hides the problem, but does not solve it.

3) The strong dependence of the optimal map S* on the initial one is due [21] to the usage of eqn. (6). This equation is derived under the unrealistic assumption (which

contradicts the aim of the MTD method) that any mono-substitution within the receptor map S leaves the equation (5) unchanged. And, because the first few correlations calculated within the MTD procedure are generally statistically irrelevant, the resulted b and $\hat{Y}$ values used in eqn. (6) are unreliable. Accordingly, the vertex assignments are questionable.

4) One can prove [21] that the MTD method should be considered a variant of the Free-Wilson procedure. Consider the Free-Wilson-type equation:

$$\hat{Y}_i = a_0 + \sum_{j=1}^{m} b_j x_{ij} \tag{7}$$

where $x_{ij} = \delta_{ij} = 1$ if the vertex j of $\hat{H}$ is occupied by a non-hydrogen atom of the biomolecule i, and $x_{ij} = 0$ if it is not occupied. Assuming that the m_1 beneficial vertices (i.e., $b_j > 0$, $j = 1 \div m_1$) and the m_2 detrimental vertices (i.e., $b_j < 0$, $j = m_1 + 1 \div m_2$) have the same weight, namely:

$$b_1 = b_2 = \ldots = b_{m_1} = b_0 = -b_{m_1+1} = \ldots = -b_{m_2} \tag{8}$$

the equation (7) becomes:

$$\hat{Y}_i = a_0 + b_0 \left[\sum_{j=1}^{m_1} x_{ij} - \sum_{j=m_1+1}^{m_2} x_{ij} \right] \tag{9}$$

One may rewrite the equation (5) as:

$$\hat{Y}_i = a + bMTD_i = a + b\left[s + \sum_{j=1}^{m} \varepsilon_j x_{ij}\right] = a + b\left[s + \sum_{j=m_1+1}^{m_2} x_{ij} - \sum_{j=1}^{m_1} x_{ij}\right]$$
$$= a' + b'\left[\sum_{j=1}^{m_1} x_{ij} - \sum_{j=m_1+1}^{m_2} x_{ij}\right] \tag{10}$$

(a_0, b_0 and a′, b′ are scaling constants). The equations (9) and (10) are formally equivalent. Because eqn. (9) was obtained under the severe assumption (8), one may not expect the conclusions of the Free-Wilson and MTD models will coincide.

The following examples illustrate the above discussion. The cardiotoxic aglycones inhibit the sodium-potassium dependent ATP-ase pump, their activity being conditioned [44] by the aglyconic moiety shown in Fig. 2.

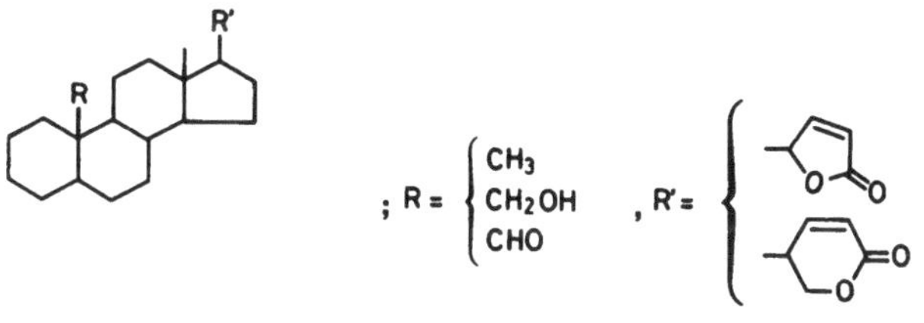

Fig. 2. Cardiotoxic aglycones: the aglyconic moiety

Recently, Simon et al. generated [22] a structure-cardiotoxic equation using a data basis consiting of 30 cardiotoxic aglycones and the hypermolecule depicted in Fig. 3.

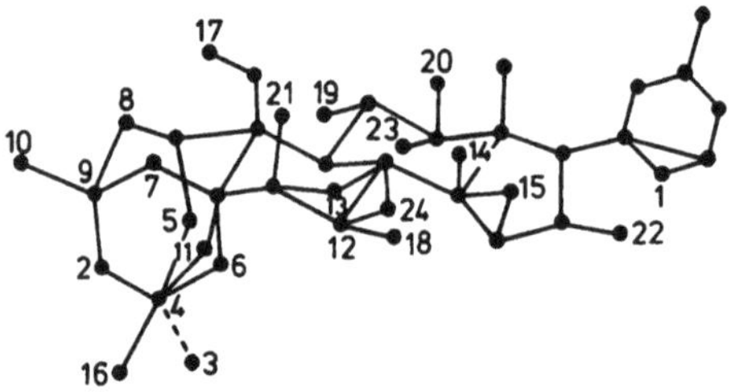

Fig. 3. Cardiotoxic aglycones: the hypermolecule

Their equation reads as (Y denotes the molar toxicity):

$$Y = 0.70\text{–}1.15\ \mathrm{MTD} \quad (11)$$
$$(r = 0.89,\ s = 0.38)$$

and corresponds to the following optimal map[1]:

$$S^* = \begin{cases} 1, 12, 14 & \text{(cavity vertices)} \\ 3, 21 & \text{(wall vertices)} \\ 2, 4\ 11, 13, 15\ 20, 22\ 24 & \text{(irrelevant vertices)} \end{cases}$$

Because the equation (11) does not improve including the hydrophobic interactions, the authors concluded that cardiotoxicity is conditioned only by steric effects. This statement contradicts the experimental finding [44] that the introduction of hydroxyl groups into steroid skeleton increases cardiotoxicity (observe that the vertex 14, occupied by the hydrophilic OH group, is classified as cavity vertex). These argue our statement that the optimal MTD values are contaminated, i.e., in the present case they code globally both steric and hydrophobic factors.

The following calculations illustrate the dependence of the optimal map, S*, on the starting one, S^0. We considered, by turn, five aglycones as delineating the $\varepsilon_j = -1$ vertices within the starting map (the other vertices were assigned as $\varepsilon_j = +1$ ones).

---

[1] The equation (11) was derived [45] by hand calculation and did not follow closely the MTD algorithm. The present author resumed [21] the calculation by computer and obtained quite different results:

$$Y = 1.462 - 0.870\ \mathrm{MTD} \qquad (r = 0.892) \quad (11a)$$

and,

$$S^* = \begin{cases} 1, 10, 12, 14 & \text{(cavity vertices)} \\ 3, 9, 13, 21, 24 & \text{(wall vertices)} \\ 2, 4, 5\text{–}8, 11, 15\text{–}20, 22, 23 & \text{(irrelevant vertices)} \end{cases}$$

The results [21] are summarized below:

a) Scilliglaucosidin; 3, 14, 10-CHO, 4Δ
$Y = 3.792 - 0.559\ \mathrm{MTD}\ (r_i = 0.506, r_{opt} = 0.914)$

$$S^* = \begin{cases} 1, 2, 5, 8, 10, 12, 17, 19, 20 & \text{(cavity vertices)} \\ 3, 6, 7, 13, 14, 21 & \text{(wall vertices)} \\ 4, 9, 11, 15, 16, 18, 22, 23, 24 & \text{(irrelevant vertices)} \end{cases}$$

b) Arenobufagin; 3, 14, 11α-triOH, 12 keto
$Y = 4.257 - 0.593\ \mathrm{MTD}\ (r_i = 0.309, r_{opt} = 0.912)$

$$S^* = \begin{cases} 1, 2, 4, 5, 8, 10, 12, 17, 19, 20 & \text{(cavity vertices)} \\ 3, 6, 7, 11, 13, 14, 15, 21 & \text{(wall vertices)} \\ 9, 16, 18, 22, 23, 24 & \text{(irrelevant vertices)} \end{cases}$$

c) Telocinobufagin; 3, 14, 11α-triOH
$Y = 4.257 - 0.593\ \mathrm{MTD}\ (r_i = 0.294, r_{opt} = 0.912)$

$$S^* = \begin{cases} 1, 2, 4, 5, 8, 10, 12, 14, 15, 17, 19, 20 & \text{(cavity vertices)} \\ 3, 6, 7, 11, 13, 21 & \text{(wall vertices)} \\ 9, 16, 18, 22, 23, 24 & \text{(irrelevant vertices)} \end{cases}$$

d) 3-deH-Scilliglaucosidin; 14-OH, 3-keto, 10-CHO, 4Δ
$Y = 4.192 - 0.592\ \mathrm{MTD}\ (r_i = 0.423, r_{opt} = 0.914)$

$$S^* = \begin{cases} 1, 2, 5, 8, 10, 12, 17, 19, 20, 23 & \text{(cavity vertices)} \\ 3, 6, 7, 13, 14, 21 & \text{(wall vertices)} \\ 4, 9, 11, 15, 16, 18, 22, 24 & \text{(irrelevant vertices)} \end{cases}$$

e) Oubagenin; 3, 14, 1, 5, 11α-pentaOH, 10-CH_2OH
$Y = 4.267 - 0.600\ \mathrm{MTD}\ (r_i = 0.134, r_{opt} = 0.916)$

$$S^* = \begin{cases} 1, 2, 4, 5, 8, 10, 11, 12, 19, 20 & \text{(cavity vertices)} \\ 3, 6, 13, 14, 15, 16, 21 & \text{(wall vertices)} \\ 7, 9, 17, 18, 22, 23, 24 & \text{(irrelevant vertices)} \end{cases}$$

r_i characterizes the MTD equations calculated against S^0, and r_{opt} against S^*. One notes that, in general, the r_i values are statistically insignificant.[2] Accordingly, the coefficient with MTD in the first iteration (see equation 5) does not differ significantly of zero, and due to equation (6) the optimization procedure cannot actually start. Relying on uncertain values of the regression coefficients to start the optimization procedure, the results will depend on the initial guess, and, accordingly, the final classification of the hypermolecule vertices is uncertain.

[2] The only r_i value slightly above the critical value of correlation coefficient corresponds to the scilliglaucosidin as initial standard ($r_{critical} = 0.463$, according to Ref. [46]).

An improved version of the MTD approach would be of real interest as a mono-parametric Free-Wilson-type method (due to their meaning, the MTD and Free-Wilson parameters belong to the same class). The topological description of the molecular structure assures the "easy to use" character of the MTD method, and the "hypermolecule" concept allows to study widely differing structures within the data basis.

2 Character and Locator Variables

Purcell and Testa [27,28] formulated a sector model as: the molecule is viewed as having regions in space (sectors) which may be either filled or unfilled by atoms or groups. The volume of the atom or group in a specified sector i defines the character variable V_i. The V_i values are estimated using the method developed by Bondi. [31]. Locator variables are defined either as geometric distances between certain atoms in molecule, or as cartesian coordinates defining the sectors. For example, using a quadrant representation [28], the affinity constant (AC) for carbonic anhydrase [29] of 34 sulfonamides follows equations such as:

$$\log AC = 0.0353\, V_1 - 0.0198\, V_3 + 0.0129\, V_4 + 6.47 \quad (12)$$
$$(r^2 = 0.828,\ s = 0.570,\ F = 48.2)$$

$$\log AC = 0.0288\, V_S + 0.424\, X - 0.544\, Y + 4.28 \quad (13)$$
$$(r^2 = 0.931,\ s = 0.362,\ F = 134.0)$$

or

$$\log AC = 0.03\, V_S + 1.08\, DIST \quad (14)$$
$$(r^2 = 0.995,\ s = 0.554,\ F = 3097)$$

where V_S is the van der Waals volume of the substituent (in mL $mole^{-1}$), V_i is the volume (in mL $mole^{-1}$) of the atom or group in the quadrant i, X and Y (in Å) stand for the cartesian coordinates defining the quadrants, and DIST represents the geometrical distance (in Å) between the sulfonamide S-atom and the substituent C-atom adjacent to the cyclic moiety.

The character variables (i.e., the group volumes) are steric parameters of known utility in QSAR, but, despite the empirical success (i.e., eqs. 13 and 14), the steric meaning of locator variables is generally rather obscure.

Motoc et al. [30] presented recently a Monte Carlo method designed to perform the sector partition of the molecular van der Waals envelope. Suggesting an heuristic approach for determining the proper intersection point of the three orthogonal planes defining the octants, this method may be regarded as a generalization of the character variables of Testa and Purcell.

Table 2. Linear interdependences between lipophilic and steric substituent constants (17 hydrocarbon substituents)

	π	f	[P]	χ	MV	MR	MW	L	B_4	B_1	E_S	ν
π	1.000	0.946	0.993	0.960	0.973	0.971	0.981	0.960	0.733	—0.194	—0.097	0.078
f		1.000	0.962	0.946	0.988	0.880	0.984	0.899	0.835	—0.048	—0.027	0.014
[P]			1.000	0.968	0.985	0.972	0.981	0.934	0.726	—0.094	—0.146	0.128
χ				1.000	0.942	0.959	0.955	0.898	0.694	—0.045	—0.116	0.089
MV					1.000	0.916	0.932	0.924	0.814	—0.090	—0.410	0.068
MR						1.000	0.998	0.900	0.580	—0.092	—0.223	0.199
MW							1.000	0.914	0.608	—0.107	—0.216	0.194
L								1.000	0.757	—0.350	—0.051	0.026
B_4									1.000	—0.267	0.123	—0.134
B_1										1.000	—0.410	0.423
E_S											1.000	—0.996
ν												1.000

Table 3. Linear interdependences between lipophilic and steric substituent constants (13 polar substituents)

	π	f	[P]	MV	MW	B_4	MR	L	B_1	χ	E_S	ν
π	1.000	0.931	0.585	0.606	0.534	0.514	0.719	0.242	0.162	0.690	—0.346	—0.043
f		1.000	0.308	0.340	0.239	0.211	0.546	0.013	0.186	0.776	—0.149	—0.043
[P]			1.000	0.978	0.970	0.966	0.809	0.763	0.201	0.153	—0.716	—0.216
MV				1.000	0.953	0.954	0.737	0.716	0.090	0.119	—0.570	—0.161
MW					1.000	0.945	0.750	0.730	0.089	0.024	—0.632	—0.162
B_4						1.000	0.735	0.668	0.057	0.037	—0.662	—0.248
MR							1.000	0.675	0.547	0.622	—0.716	—0.277
L								1.000	0.463	0.144	—0.533	—0.240
B_1									1.000	0.625	—0.497	—0.039
χ										1.000	—0.259	—0.008
E_S											1.000	0.201
ν												1.000

3 Possible Connections Between Steric and Other Substituent Constants

There exist several indications of correlations between hydrophobicity or solubility in water and molecular volume or surface, parachlor, molar refraction, or molecular connectivity index [32-40].

Recently, Tichy investigated [41] the dependencies of the steric constants, E_S, ν, L, B_1, B_4, MV (molar volume), [P] (parachor), MR (molar refraction), MW (molecular weight), and χ (molecular connectivity index) on lipophilicity, as it is measured by π [42] and f [43] constants. The data were treated by factor analysis methods.

Examining the results collected in Tables 2 and 3, this author concluded that E_S, ν and B_1 have the greatest chance to form an orthogonal class with the lipophilic constants π and f.

4 References

1. Hopfinger, A. J.: J. Am. Chem. Soc. *102*, 7196 (1980); Arch. Biochem. Biophys. *206*, 153 (1981); Battershell, C., Malhorta, D., Hopfinger, A. J.: J. Med. Chem. *24*, 812 (1981)
2. Motoc, I.: Arzneim. Forsch./Drug Res. *31*, 290 (1981); Motoc, I., Dragomir, O.: Math. Chem. *12*, 117, 132 (1981); Motoc, I., Muscutariu, I., Dragomir, O.: Kexue Tongbao *27*, 1333 (1982); Motoc, I., Reilly, P. M.: Math. Chem. *13*, 333 (1982)
3. Amoore, J. E.: Ann. N.Y. Acad. Sci. *116*, 457 (1964)
4. Amoore, J. E., Palmieri, G., Wanke, E.: Nature *216*, 1084 (1967)
5. Amoore, J. E.: Nature *214*, 1095 (1967)
6. Ruzicka, L.: Chem. Ztg. *44*, 129 (1920)
7. Amoore, J. E. et al.: Science *165*, 1266 (1969)
8. Amoore, J. E.: Molecular Basis of Odor, Springfield (Ill.), Thomas, 1970
9. Motoc, I. et al.: Can. J. Pharm. Sci. *14*, 96 (1979)
10. Amoore, J. E., Venstrom, D.: In: Proc. 2-nd Int. Symp. Olfaction and Taste, Tokyo, 1965
11. Allinger, N. L.: In: Pharmacology and the Future of Man, vol. 5, p. 57, Proc. 5-th Int. Congr. Pharmacol., 1972
12. Motoc, I., Muscutariu, I.: Math. Chem. *8*, 367 (1980)
13. Shreider, Yu. A. (ed.): The Monte Carlo Method, London, Pergamon, 1966; Bruns, W., Motoc, I., O'Driscoll, K. F.: Monte Carlo Applications in Polymer Science, ch. 1, Lecture Notes in Chemistry, vol. 25, Berlin, Springer, 1982
14. Motoc, I.: Math. Chem. *4*, 113 (1978)
15. Motoc, I. et al.: Studia Biophys. (Berlin) *66*, 75 (1977)
16. Balaban, A. T., Chiriac, A., Motoc, I., Simon, Z.: Steric Fit in QSAR, Lecture Notes in Chemistry, vol. 15, Berlin, Springer, 1980; a) ch. 4; b) ch. 6; c) ch. 5; d) Appendix; e) Simon, Z., Szabadai, Z.: Studia Biophys. (Berlin) *39*, 123 (1973)
17. Simon, Z. et al.: Studia Biophys. (Berlin) *55*, 217 (1976)
18. Simon, Z. et al.: Studia Biophys. (Berlin) *59*, 181 (1976)
19. Simon, Z., Badilescu, I. I., Racovitan, T.: J. Theor. Biol. *66*, 485 (1977)
20. Motoc, I.: Math. Chem. *5*, 275 (1979)
21. Motoc, I.: unpublished
22. Simon, Z. et al.: Eur. J. Med. Chem. *15*, 521 (1980)
23. Wilson, K.: 144-th Natl. Amer. Chem. Soc. Met., Spring 1963, Div. Chem. Ed. 23
24. Kaufmann, J. J.: Int. J. Quant. Chem. *QBS4*, 375 (1977)
25. a) Niculescu-Duvaz, I. et al.: Carcinogenesis *2*, 269 (1981); b) Verloop, A.: Phil. Trans. R. Soc. Lond. *B295*, 45 (1981)
26. Carles, P., Goursot, M.: Symp. Methods Avancées QSAR, Paris, March 16–18, 1979

27. Purcell, W. P., Testa, B.: In: Biological Activity and Chemical Structure, (ed. Keverling Buisman, J. A.) p. 269, Amsterdam, Elsevier, 1977
28. Testa, B., Purcell, W. P.: Eur. J. Med. Chem. *13*, 509 (1978)
29. King, R. W., Burgen, A. S. V.: Proc. Roy. Soc. Lond. *B193*, 107 (1976)
30. Motoc, I., Dragomir, O., Muscutariu, I.: Math. Chem. *8*, 323 (1980)
31. Bondi, A.: J. Phys. Chem. *68*, 441 (1964)
32. Amidon, G. L., Yalkowski, S. H., Leung, S.: J. Pharm. Sci. *63*, 1858 (1974)
33. Craig, P. N.: J. Med. Chem. *14*, 680 (1971)
34. Harris, M. J., Higuchi, T., Rytting, J. H.: J. Phys. Chem. *77*, 2694 (1973)
35. Hermann, R. B.: J. Phys., Chem. *77*, 2754 (1972)
36. Leo, A., Hansch, C., Low, P. Y. C.: J. Med. Chem. *19*, 611 (1976)
37. Reynolds, J. A., Gilbert, D. B. Tanford, C.: Proc. Natl. Acad. Sci. US *71*, 2925 (1974)
38. Amidon, G. L., Anik, S. T.: J. Pharm. Sci. *65*, 801 (1976)
39. Hall, L. H., Kier, L. B., Murray, W. T.: J. Pharm. Sci. *64*, 1974 (1975)
40. Murray, W. J., Hall, L. H., Kier, L. B.: J. Pharm. Sci. *64*, 1978 (1975)
41. Tichý, M.: Int. J. Quant. Chem. *16*, 509 (1979)
42. Fujita, T., Iwasa, J., Hansch, C.: J. Am. Chem. Soc. *86*, 5175 (1964)
43. Rekker, R. F.: The Hydrophobic Fragmental Constants, Amsterdam, Elsevier, 1977
44. Baumgarten, G. (ed.): Die Herzwirksamen Glykoside, p. 199–225, Leipzig, VEB G. Thieme, 1963
45. Simon, Z.: personal communication
46. Crow, E. L., Davis, F. A., Maxfield, M. W.: Statistic Manual, p. 241, New York, Dover Publications, 1960

Volume and Bulk Parameters

Marvin Charton

Chemistry Department, Pratt Institute, Brooklyn, NY 11205/USA

Table of Contents

1 Introduction

Thanks largely to the work of Hansch and his collaborators, the general relationship used in the application of correlation analysis to bioactivities is of the form

$$BA + L\sigma_{IX} + D\sigma_{DX} + S\upsilon_X + T_1\tau_X + T_2\tau_X^2 + B_0 \quad (1)$$

in which:

σ_I is the localized (field/inductive) electrical effect parameter
σ_D is the delocalized (resonance) electrical effect parameter
υ is the steric parameter
τ is the transport parameter.

Neither the transport parameter nor the steric parameter was well understood in biological systems. In view of the partical importance of quantitative structure activity relationships (QSAR) for the design of bioactive molecultes it is not surprising that an intensive search for useful physicochemical parameters which would give a better representation of steric and transport contributions has been carried out. It sometimes seems as though almost every known physical property or substituent constant has been suggested as a parameter for improved QSAR. Among the more popular examples of this trend are the so-called bulk parameters: group volume, group area and group mass.

1.1 "Bulk" parameters

We define a group volume parameter as one whose dimensions are volume (l^3). Examples include the van der Waals volume, V_W [1]; molar refractivity, MR [2]; Exner molar volume, V_M [3]; Traube's rule volume, V_T [4]; molal volume, $\bar{V}^0_{25}$ [5]; and Parachor, P [6]. The group surface area parameter is defined as one whose dimensions are area (l^2). The van der Waals area, A_W, is an example. The mass parameter has the dimension m, the only example is the group weight, W_M. Examples of correlations with volume, area and mass parameters are reported in Table 1. That the use of these parameters frequently results in significant correlations is undoubtedly true. If the only objective in modelling bioactivities was the prediction of the activity of new compounds there would be no pressing reason for going further. If the design of bioactive substances is to be transformed from an art into a science then there is every reason for going further. What we must be able to do in order to effect that transformation is to interpret the meaning of correlations with bulk parameters. In the past interpretation has been both vague and varied. Some authors have proposed that group volume parameters are a measure of some kind of steric effect. Others have proposed that they represent group hydrophobicity and are therefore transport parameters. To clarify the nature of these properties we must:

1. Show the relationships between the different types of bulk parameter.
2. Describe the nature of steric effects in biological systems.
3. Describe the nature of transport in biological systems.

Table 1. Correlations with Volume, Surface, Molecular Weight Parameters

Parameter (Interpretation)	Substrate	Biocomponent	Q	Ref.
V_W(S)	4-nitrophenyl-2-acyl-amino-2-deoxy-β-D-glucopyranosides	Taka-N-acetyl-β-D-glucosaminidase	$\log V_{rel}$	9)
V_W(H)	alcohols, esters, etc.	frog muscle (narcosis)	pC	10)
V_W(H)	a)			11)
V_W(S)	2-substituted isonicotinic acid hydrazides	bacteria	pMIC	12)
V_W, V_W^2(S)	trimethoprim acid analogs	bacteria	pA	13)
ΔV_W, ΔV_W^2, ΔMR, ΔMR^2(S)	lindane analogs	insects	pLD_{50}	14)
$\log V_W$, $(\log V_W)^2$, (S, B)	aryloxypropionic acid	flax root (growth rate inhibition)		15)
$\log V_W$, $(\log V_W)^2$, (S, B)	quarternary ammonium compounds	guinea pig ileum (acetylcholine receptor)	$\log K_1$ pA_2	16)
$\log V_W$, $(\log V_W)^2$, (S, b)	imidazolines	isolated rabbit intestine (α-adrenergic agonistic activity)	pD	17)
V_T	8-substituted-10-(4-methyl-piperazino)-10,11-dihydro-dibenzo(b,f)thiepin	mice (rotating rod test)	pED_{50}	4)
$\bar{V}_{25°}$ (IMF)	2-substituted ethyl acetates	acetylcholinesterase	$\log\left(\frac{k_{2n}}{K_S}\right)$	18
P(H)	a)			5)
ΔA_W(H)	androst-4-en-3-one derivatives	rabbit	$\log A_{rel}$	6)
W_M(none)	N-alkyl-N-ethylpiperidinium bromides	mice	LD_{50}	7)
W_M(none)		human (odor threshold conc.)	log C	19)

Correlations with MR are so numerous that examples are generally not given here. Abbreviations: S, steric; B, bulk; H, hydrophobicity; IMF, intermolecular forces.
a) Various examples are given in the reference cited.

2 The Nature of the Steric and Transport Parameters

Though many authors who have used group volume, group area and group mass parameters are vague in their interpretation of the significance of these parameters, there seems to be general agreement that they represent either a steric effect, or a hydrophobicity (transport) effect, or some combination of the two. If we are to arrive at an understanding of the validity of correlations with these parameters and to interpret them, it is necessary for us to examine the nature of steric and transport effects in biological systems.

2.1 Biological Systems

We may define the biological system undergoing study as a biocomponent (e.g. enzyme, cell, tissue, organism) which is interacting with a bioactive substance. In the design of bioactive substances it is necessary to study a wide variety of biological systems. They range from pure enzymes through microsomes, cell homogenates, whole cells, tissue samples and organs to large multicellular organisms, such as mice, rats and humans. If we use the MacFarland model of bioactivity as a basis for discussion then the mode of action of a bioactive substance may consist of the following steps:

1. Transport of the bioactive substance (bas) from the site of introduction to some receptor site on a biopolymer.
2. a) Recognition;
   b) bas-receptor complex formation.
3. Chemical reaction.

Bioactivity may result directly from step 2b or from step 3. In the case of bioactivity in a mammal transport may involve crossing of a number of biomembranes. Transport is much simpler in the case of single cells. In the study of pure enzymes, step 1 vanishes. Obviously, interpretation of a QSAR will depend on the type of biocomponent in the biological system which is studied. As the pure enzyme is the simplest system, it is the easiest to interpret.

2.2 Steric Effects in Biological Systems

Steric effects result from repulsions between valence electrons or non-bonded atoms. Steric effects always increase the energy of a chemical species in which they are present. The overall steric effect on a chemical reaction may be either favorable or unfavorable. If steric effects in the reactant are larger then in the product (or transition state) then the reaction is favored (steric augmentation). If the reverse is the case the reaction is disfavored (steric diminution). This is also true of a dynamic physical property involving initial and final states, such as ionization potential. We may expect the same result in biological systems for the formation of the bas-receptor complex, and when it occurs, for the subsequent chemical reaction of the complex.

A considerable body of evidence has appeared which indicates that the receptor site in an enzyme may be located in a very flexible pocket, and that protein molecules in general can show a startlingly large degree of conformational flexibility. This does not affect the nature of steric effects in biological systems. Excluding steric effects in the transport step of bioactivity (2.1) we may divide the steric effect into two components:

a) Steric effects involving the chemical reaction in step 3.
b) Steric effects involved in the recognition of the bas and the formation of the bas-receptor complex.

The steric effects in a are directly comparable to those observed for ordinary chemical reactivities. They involve only groups in proximity to those atoms which are actually involved in bond making and breaking. The steric effects in b can involve any group of atoms in the bas which is in van der Waals contact with the receptor or with the biopolymer on which the receptor is located. If the receptor site lies in a pocket which can adjust to fit any bas no matter what its size or shape then *no steric effect will be observed.* If, however, the parent biopolymer has limited conformational flexibility, and, as is likely, this flexibility is not the same in all directions, then a steric effect will be observed. Furthermore, the steric effect will be conformationally dependent, and it is probable that the minimal steric interaction (MSI) principle will be observed. The MSI principle states that a substituent whose steric effect is conformationally dependant will prefer that conformation which minimizes steric repulsions and therefore will give rise to the smallest steric effect. It follows, then, that steric repulsion effects in biological activity are not fundamentally different from those on chemical reactivity or physical properties.

2.3 Transport Effects in Biological Systems

When a bioactive substance is introduced into a multicellular organism it must travel from the point of entry to some receptor site by which it is recognized and to which it becomes bound. The trip to the receptor may involve the crossing of a number of biomembranes. In order to cross a biomembrane the bas must generally be transferred from an aqueous medium to one which is very much more lipophilic. In the aqueous medium the bas is surrounded by a cage of water molecules attracted to it by intermolecular forces. When the bas has migrated to the surface of the biomembrane it is bound to it by other intermolecular forces. If the process reaches equilibrium, the equilibrium constant resembles a partition coefficient, the bas having been "partitioned" between the aqueous phase and the biomembrane. The magnitude of ΔG for this process will be a function of the difference in intermolecular forces (imf) for the two systems:

1. aqueous phase + bas
2. membrane + bas

or

$$\Delta G = f(imf_2 - imf_1) \tag{2}$$

When a bas binds to a receptor site the process is analogous. Thus $\Delta G'$ for binding will be a function of the Δimf for

1. aqueous phase + bas
3. receptor + bas

and

$$\Delta G' = f\,\Delta imf_{3,1} = f(imf_3 - imf_1) \tag{3}$$

The function of a transport parameter is to model the transfer of the bas from the aqueous phase to biomembrane and bas receptor. The transport parameter is frequently also referred to as a hydrophobicity or lipophilicity parameter, the former term is no doubt preferred by pessimists and the latter by optimists. Unfortunately, there has been no attempt at the standardization of nomenclature in this field (A rose by any other name . . .). As is usually the case under these circumstances far too much heat and very little light results.

From equations 2 and 3 it follows that the transport phenomena is a composite parameter, representing a number of different contributions to the imf. The validity of these equations is supported by the reports of Charton and Charton [20], who have shown that hydrophobicity parameters, log P values, and Rekker [21] fragmental constants for amino acids with side chain X are very well correlated by the equation

$$Q_X = A\alpha_X + L\sigma_{IX} + S\upsilon_X + H_1 n_{HX} + H_2 n_{nX} + I i_X + B_0 \tag{4}$$

or relationships derived from it. In this equation α is the polarizability parameter; σ_I the localized electrical effect parameter; υ the steric parameter, n_H is equal to the number of OH or NH bonds in X, n_n is equal to the number of valence shell lone pairs in X, and i = 0 for uncharged and 1 for charged X. The imf which have been considered are (parameters which model them are in brackets)

1. Hydrogen bonding (hb) [n_H, n_n]
2. Ion—dipole (Id) [i]
3. Ion—induced dipole (Ic) [i]
4. Dipole—Dipole (dd) [σ_I]
5. Dipole—Induced dipole (di) [α, σ_I]
6. Induced dipole—Induced dipole (ii) [α]

The steric parameter accounts for steric effects on the solvation of α-amino and carboxyl groups caused by the side chain. We have not attempted to account for any charge transfer interactions as none of the compounds studied is a very effective charge transfer acceptor, and the water and n-octanol phases can only function as change transfer donors. Further support is provided by the correlation of log P values for $Ph(CH_2)_mX$ (m = 1, 2, 3) with the equation

$$Q_X = A\alpha_X + L\sigma_I + S\upsilon_X + H_1 n_{HX} + H_2 n_{nX} + B_0 \tag{5}$$

Table 2. Transport Parameters Correlated with Equation 5
LP 1, 2, 3 log P values for $Ph(CH_2)_nX$ with n = 1, 2, 3

n/X	H	F	Cl	Br	OH	NH_2	CO_2Me	CO_2H
1	2.69	—	—	—	1.10	1.09	1.83	1.41
2	3.15	—	2.95	3.09	1.36	1.41	2.32	1.84
3	3.68	2.95	3.55	3.72	1.88	1.83	2.77	2.42
n/X	**CN**	**Ac**	**$CONH_2$**	**OAc**	**OMe**	**NMe_2**	**Me**	
1	1.56	1.44	0.45	1.96	—	—	3.15	
2	1.72	—	0.91	2.30	—	—	3.68	
3	2.21	2.42	1.41	2.77	2.70	2.73	—	

LP 4 log P values for PhX
NH_2, 0.90; OMe, 2.11; H, 2.13; Cl, 2.84; CO_2Et, 2.62; NO_2, 1.85; OH, 1.46; C_2H, 2.53; CH_2CONH_2, 0.45; CH_2OH, 1.10
P1 π aliphatic values
OH, −1.16; OMe, −0.47; SMe, 0.45; F, −0.17; Cl, 0.39; Br, 0.60; I, 1.00; NH_2, −1.19; NHMe, −0.67; NMe_2, −0.32; Me, 0.50; C_2H, 0.48; $CONH_2$, −1.71; Ac, −0.71; CN, −0.84; CO_2Me, −0.27; OAc, −0.27; NO_2, −0.85; H, 0.
Data are from Ref. [22].

Table 3. Results of Correlations with Eq. 5

LP 1	$\log P_X = -2.30(\pm 6.28)\alpha_X - 2.59(\pm 0.864)\sigma_{IX} + 1.07(\pm 1.29)\upsilon_X$ $-0.728(\pm 0.173)n_{HX} - 0.0608(\pm 0.116)n_{nX} + 2.59(\pm 0.364)$
	S = 0.382 $100R^2 = 89.60$ F = 6.893 n = 10
LP 2	$\log P_X = -0.700(\pm 5.82)\alpha_X - 2.77(\pm 0.744)\sigma_{IX} + 2.01(\pm 1.05)\upsilon_X$ $-0.820(\pm 0.158)n_{HX} - 0.159(\pm 0.100)n_{nX} + 2.91(\pm 0.323)$
	S = 0.354 $100R^2 = 91.96$ F = 11.44 n = 11
LP 3	$\log P_X = -11.1(\pm 4.65)\alpha_X - 3.27(\pm 1.21)\sigma_{IX} + 3.92(\pm 1.50)\upsilon_X$ $-0.932(\pm 0.185)n_{HX} - 0.111(\pm 0.0874)n_{nX} + 3.38(\pm 0.332)$
	S = 0.384 $100R^2 = 81.15$ F = 6.889 n = 14
LP 4	$\log P_X = 8.91(\pm 4.62)\alpha_X + 1.20(\pm 0.766)\sigma_{IX} - 1.08(\pm 989)\upsilon_X$ $-0.620(\pm 0.188)n_{HX} - 0.281(\pm 0.108)n_{nX} + 2.22(\pm 0.258)$
	S = 0.287 $100R^2 = 94.25$ F = 13.12 n = 10
P1	$\pi_X = -3.64(\pm 3.11)\alpha_X - 1.10(0.522)\sigma_{IX} + 2.46(\pm 0.771)\upsilon_X$ $-0.695(\pm 0.130)n_{HX} - 0.145(\pm 0.0655)n_{nX} - 0.374(\pm 0.254)$
	S = 0.327 $100R^2 = 85.07$ F = 14.81 n = 19

The data used are presented in Table 2, and the results of the correlations in Table 3. The steric term was included to account for the observation that conformation in molecules of the type XCH_2CH_2Z where Z is constant is a function of both σ_{IX} and υ_X. Unfortunately as is shown by the partial correlation coefficients, υ_X is significantly linear in α_X. The results nevertheless support the general validity of equation 4. Further support is provided by the correlation of aliphatic π_X values with equation 5. Again, the data used and the results of the correlations are presented in Tables 2 and 3 respectively.

3 Relationships Among Bulk Parameters

In order to properly understand the use and significance of bulk parameters it is necessary to consider the interrelationships among them.

3.1 Relationships Among Volume Parameters, Parachor and Molar Refractivity

It has been shown [8] that V_T, the molar volume calculated from Traube's rule [4], V_M, the molar volume of Exner [3], and the parachor, [5] P all give significant correlation with the equation

$$Q_X = a_1 V_{WX} + a_0 \tag{6}$$

where V_W is the Bondi van der Waals volume [1]. Results of the correlations are set forth in Table 4 (sets 2, 3, 4). MR_X values also give significant correlation with Eq. 6 (set 10, Table 4) although best results are obtained when separate categories of substituent are correlated. Correlations are reported in Table 4 for ZH_n groups ($n = 0$–2) (set 1A); alkyl σAk) groups (set 1B), π-bonded groups (set 1C), and partially π-bonded groups (set 1D). As V_M, V_T, and MR are all correlated by V_W, they must be related to each other. They are obviously parameterizing the same factor.

Table 4. Correlations with Equation 6

Set	a_1	a_0	r[a]	F[b]	S_{est}[c]	S_{a_1}[c]	S_{a_0}[c]	n[d]	$100r^2$[e]
1	0.714	− 2.49	0.968	51.62[f]	1.72	0.0994[f]	1.39[g]	13	82.45
1A	0.968	− 4.94	0.989	269.7[f]	0.712	0.0589[f]	0.794[f]	8	97.82
1B	0.442	− 0.197	0.9994	7178[f]	0.230	0.00522[f]	0.217[j]	11	99.87
1C	0.564	− 1.58	0.999	4130[f]	0.634	0.00878[f]	0.373[i]	13	99.73
1D	0.546	− 2.85	0.992	522.1[f]	0.767	0.0239[f]	0.831[i]	10	98.49
2	1.27	6.80	0.873	31.90[f]	3.83	0.225[f]	3.10[h]	12	76.22
3	1.60	− 4.14	0.993	1064[f]	2.56	0.0490[f]	1.61[i]	17	98.61
4	4.05	2.83	0.996	896.5[f]	3.69	0.135[f]	2.28[j]	10	99.12
5	4.06	−16.9	0.611	6.566[k]	27.5	1.59[k]	22.1[j]	13	37.38
6	2.45	− 5.95	0.997	1958[f]	2.59	0.0553[f]	1.20[f]	13	99.44
7	0.218	3.10	0.751	14.22[l]	2.71	0.0578[i]	1.25[k]	13	56.39

a. Correlation coefficient. b. F test. Superscript indicates the confidence level (CL). c. Standard errors of the estimate, m and c. Superscripts indicate CL of "Student t" test. d. Number of points in the set. e. The percent of the data accounted for by the regression equation. f. 99.9% CL. g. 80.0% CL. h. 90.0% CL. i. 99.0% CL. j. 50.0% CL. k. 95.0% CL. l. 99.5% CL.

3.2 Molecular Weight

Molecular weight, W_M, is not a linear function of V_W (set 5, Table 4). It is, not surprisingly, well correlated by the equation

$$Q_X = a_{11}n_{e,X} + a_{10} \tag{7}$$

where n_e is the total number of electrons in X (set 6, Table 4). MR gives a poor but significant correlation with Eq. 7 (set 7, Table 4). Much better results are obtained with the equation

$$MR_X = a_b n_{bX} + a_c n_{cX} + a_n n_{nX} + a_0 \tag{8}$$

where n_b, n_c, and n_n are the numbers of bonding —, core- and non-bonding electrons respectively in X. For 53 points the value of $100R^2$ obtained is 95.2, with $s = 2.10$ and $F = 323.7$. The term in n_n is of borderline significance. The coefficients a_b, a_c, a_n, and a_0 are 0.677, 0.323, —0.0805 and 1.09 respectively. As M_W and MR are both related to the number of electrons they are clearly related to each other.

3.3 Volume Parameters as a Function of Steric Parameters

Some degree of relationship between steric effect parameters and volume parameters must be expected. Consider for example monatomic substituents. A single atom, X, may be considered sperical. Its volume is therefore given by the expression for the volume of a sphere.

$$V_X = (4/3)\,\pi r_{VX}^3 \tag{9}$$

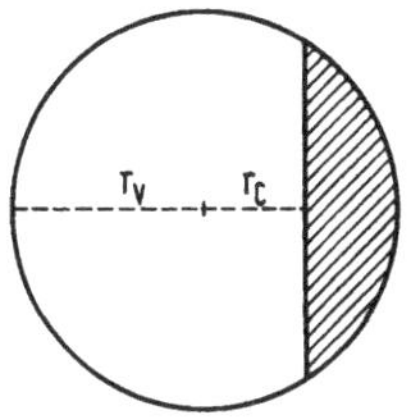

Fig. 1. Van der Waals volume of a monatomic substituent. The quantities r_V and r_C are the van der Waals and covalent radii respectively. The shaded area is cut off from the sphere when a covalent bond forms

where r_{VX} is the van der Waals radius of X. The van der Waals volume V_W of X is the volume that results when a plane perpendicular to a diameter of the sphere at a distance r_{CX} (the covalent radius of X) from the nucleus of the atom (and center of the sphere) cuts a slice off V_X (Fig. 1). As

$$r_{VX} = r_{CX} + b \tag{10}$$

and as any set of numbers is significantly linear in its cube, it follows that

$$m_i^3 = a_1 m_i + a_0 \tag{11}$$

This is indeed borne out by correlating V_W for monatomic X with r_V; for the set including H and halogen, $100r^2 = 90.0$, further inclusion of OH, SH, NH_2, and PH_2 results in $100r^2 = 95.3$. Note however, that when the cylindrical groups, CN and C≡CH are also included the $100r^2$ value falls to 85.8. To demonstrate even more clearly the sharp difference between steric effects and volume parameters we

Table 5. Values of φ_S and φ_V for Isochoric and Isosteric Sets

1. φ_S for C_4 Ak groups, $V_W = 44.35 \pm 0.01$
   Bu, 1.00; iBu, 1.44; sBu, 1.50; tBu, 1.82
2. φ_S for C_5 Ak groups, $V_W = 54.58 \pm 0.01$
   Am, 1.00; iPrCHMe, 1.90; Et_2CH, 2.22; $EtMe_2C$, 2.40; iAm, 1.00; PrMeCH, 1.54; $tBuCH_2$, 1.97; $sBuCH_2$, 1.47
3. φ_S for C_5 Ak groups, $V_W = 64.805 \pm 0.015$
   Hx, 1.07; $tBuCH_2CH_2$, 1.03; iBuMeCH, 1.60; tBuMeCH, 3.10; $iAmCH_2$, 1.00; BuMeCH, 1.57; iPrEtCH, 3.10
4. φ_S for groups with $V_W = 15.49 \pm 0.79$[a]
   CH_2OH, 1.51; Br, 1.86; CN, 1.14; PH_2, 1.71; SH, 1.71; NO_2, 1,3.97; CHO, 1.43, 2.71

11. φ_V fir iAm and $(CH_2)_nEt$ $\upsilon = 0.68$
   Pr, 1; Bu, 1.30; Am, 1.60; $iAmCH_2$, 1.90; Oc, 2.50; No, 2.80; $NoCH_2CH_2$, 3.40; $No(CH_2)_4$, 4.00; $No(CH_2)_6$, 4.60; $No(CH_2)_8$, 5.20
12. φ_V for groups with $\upsilon = 0.553 \pm 0.0226$
   CH_2NH_2, 1.73; Me, 1.14; CH_2OH, 1.30; Cl, 1; C_2H, 1.63; OPr, 3.15; OBu, 4.01; Et, 1.99; OAm, 4.86; OCH_2CH_2tBu, 5.71

a. For the planar π bonded NO_2 and CHO groups, φ_S values for both υ_{min} and υ_{max} are reported.

shall compare sets of groups which are isochoric (V_W = constant) and sets which are isosteric. In Table 5, values of the quantity, φ_S are given. They are defined by

$$\varphi_S \equiv \upsilon_X/\upsilon_{X,min} \tag{12}$$

where υ_X is the steric parameter for X and $\upsilon_{X,min}$ (the minimum value of υ) is the steric parameter for the smallest group in the set. These sets are isochoric. The quantity

$$\varphi_V \equiv V_{WX}/V_{W,min} \tag{13}$$

is given for isosteric sets. These sets clearly demonstrate that although poor choice of substituents will frequently lead to a relationship between volume and steric parameter, this need not always be the case.

3.4 Bulk Parameters as a Function of Polarizability

Molar refractivity is an additive property. If we write a bioactive species as X-G-Y where X is the variable substituent; Y, the active site responsible for the measurable bioactivity; and G, the skeletal group to which X and Y are bonded; it follows that

$$MR_{XGY} = MR_X + MR'_{GY} \tag{14}$$

Thus, the use by Hansch and his coworkers of MR_X values is justifiable. It is well known that MR is related to the polarizability, α', by the equation

$$MR = 4/3\ \pi N_A \alpha' \tag{15}$$

where N_A is Avogadro's number. MR_X is therefore a measure of the group polarizability of the substituent X. As MR_X is linear in V_{WX}, and V_T, V_M, and P are linear in V_{WX}, these quantities are also closely related to α. Though sets of MR_X, V_{TX}, V_{MX} or P_X can be designed which are independent of steric parameters such as the isosteric sets in Table 5, no such set can be designed which is independent of polarizability. The W_M parameter is also related to polarizability, although it is a much cruder measure of it than are the other bulk parameters. As A_{WX}, the van der Waals surface area of X is linear in V_{WX}, it too is a measure of α' [23].

4 The Significance of Bulk Parameters

The results lead to the conclusion that a transport parameter such as log P or π is a composite quantity and is a function of intermolecular forces. If this is the case, it follows that log P or π may not always have a suitable composition for modeling a particular case of bioactivity. The dependence of bioactivity on Δimf should vary with the nature of the membrane to be crossed and with the nature of the receptor site to which the bioactive substance is to bind. The receptor site consists of some region on a biopolymer which is characterized by a gross shape and by some number of atoms and groups of atoms which constitute its surface.

The shape can be altered to some degree by conformational changes, but only at the expense of energy. As a result, the receptor shape must be of decidely limited flexibility. To the extent to which the shape can be changed to accomodate the bioactive substance, no steric effect will be observed. When the cost in energy of altering the conformation becomes sufficiently great a true (vector) steric effect dependent on the conformation of a non-symmetric group of atoms will be observed. What, then, do "bulk" parameters actually represent? Let us sum up the evidence.

1. Steric effects in organic and inorganic chemistry are vector quantities.
2. The bioactivity of a chemical species is dependent on its physical and chemical properties and on nothing else.
3. Molar volume and those quantities which are a linear function of it are scalar quantities.
4. Molar volume and those quantities linearly related to it are proportional to polarizability.
5. Polarizability is a parameter which is a measure of induced dipole-induced dipole (also known as London or dispersion forces).
6. The quantity Δimf on which both transport and receptor binding of a bioactive substrate depend is variable in composition. It depends on the nature of the atoms or groups of atoms which make up the membrane surface or alternatively the receptor surface.

From 1, 2, and 3 we conclude that bulk parameters are *not* a true measure of steric effects. From 4, 5, and 6 we conclude that they represent ii and to some extent, di forces.

The best "bulk" parameter to use as a measure of polarizability is that for which the largest number of good experimental values is available. This seems to be group molar refractivity. As all of the other "bulk" parameters are highly linear in molar refractivity there is no advantage to their use. Of course, this will not prevent authors in desperate search of another least publishable unit from using still other quantities which are also linear in molar refractivity. A particularly worthless argument is that for the particular data set studied, the correlation coefficient is somewhat better with the new parameter than with molar refractivity. Variations in the goodness of fit for individual bioactivity data sets are to be expected, particularly in view of the "softness" of much of the data. The kindest statement that can be made about the proposal of a new parameter on the basis of a single data is that it is asinine. It is far better to temper the lust for publication with a little thought. Equally outrageous is the use of a physically absurd quantity such as the logarithm of the group molecular weight.

The results presented indicate that in the most general case the transport parameter τ should take the form

$$\tau_X = A\alpha_X + L\sigma_{IX} + S\upsilon_X + H_1 n_{HX} + H_2 n_{nX} + Ii_X + C_1\beta_{DX} + C_2\beta_{AX} + B_0 \quad (16)$$

where β_D and β_A are the charge transfer donor and acceptor parameters respectively. This must be the case if they are to successfully account for Δimf in all possible cases. We have already noted that Δimf must vary with the nature of the membrane or receptor site, and for that matter with the nature of the aqueous phase as well.

If in a particular case of interest A $\gg$ L, S, H_1, H_2, I, C_1, C_2 then bioactivity may be determined solely by a bulk parameter on the condition that transport or binding is the bioactivity-determining step. A given correlation may contain terms in both π or log P and in a bulk parameter when π or log P does not itself properly model the Δimf, and the bulk parameter may then be used to account for the difference.

In conclusion, then, when true steric effects have been eliminated by proper choice of the members of the data set the bulk parameter constitutes a measure of ii (dispersion) and di interactions involved in either transport across membranes or in the formation of a substrate — receptor complex.

5 References

1. Bondi, A.: J. Phys. Chem. *68*, 441 (1964)
2. Hansch, C., Leo, A., Unger, S. H., Kim, K. H., Nikaitani, D., Lien, E. J.: J. Med. Chem. *16*, 1207 (1973)
3. Exner, O.: Collect. Czech. Chem. Commun. *32*, 1 (1967)
4. Tollenaere, J. P., Moereels, H., Protiva, M.: Eur. J. Med. Chem. — Chim. Ther. *11*, 293 (1976)
5. Ahmad, P., Fyfe, C. A., Mellors, A.: Biochem. Pharmacol. *24* 1103 (1975)
6. Lee, D. L., Kollman, P. A., March, F. J., Wolff, M. E.: J. Med. Chem. *20*, 1139 (1977)
7. Vrbovsky, L.: Quantitative Structure Activity Relationships, Budapest, Akademiai Kiado, 1973
8. Charton, M., Charton, B. I.: J. Org. Chem. *44*, 2284 (1979)
9. Yamamoto, K.: J. Biochem. (Japan) *76*, 385 (1974)
10. Moriguchi, I., Kanada, Y., Komatsu, K.: Chem. Pharm. Bull. *24*, 1799 (1976)
11. Moriguchi, I., Kanada, Y.: Chem. Pharm. Bull. *25*, 925 (1977)
12. Seydel, J. K., Schaper, K. J., Wempe, E., Cordes, H. P.: J. Med. Chem. *19*, 483 (1976)
13. Dinya, Z., Toth-Martinez, B., Rohlitz, S.: Acta Pharm. Hungarica *46*, 1 (1976)
14. Kiso, M., Fujita, T., Kurihara, A., Uchida, M., Tanaka, K., Nakajima, M.: Pesticide Biochem. Physiol. *8*, 33 (1978)
15. Lien, E. J., Rodrigues de Miranda, J. F., Ariëns, E. J.: Mol. Pharmacol. *12*, 598 (1976)
16. Lien, E. J., Ariëns, E. J., Beld, A. J.: Eur. J. Pharmacol. *35*, 245 (1976)
17. Boudier, H. S., de Boer, J., Smeets, G., Lien, E. J., van Rossum, J.: Life Sciences *17*, 377 (1975)
18. Hasan, F. B., Cohen, S. G., Cohen, J. B.: J. Biol. Chem. *255*, 3898 (1980)
19. Andreyeshcheva, N. G.: Environ. Health Perspectives *13*, 27 (1976)
20. Charton, M., Charton, B. I.: J. Theoret. Biol. (1982)
21. Rekker, R. F.: The Hydrophobic Fragmental Constant, Amsterdam, Elsevier, 1977
22. Tute, M. S.: Adv. Drug Res. *6*, 1 (1971)
23. Leo, A., Hansch, C., Jow, P. Y. C.: J. Med. Chem. *19*, 611 (1976)

Applications of Various Steric Constants to Quantitative Analysis of Structure-Activity Relationships

Toshio Fujita and Hajime Iwamura

Department of Agricultural Chemistry, Kyoto University, Kyoto 606, Japan

Table of Contents

Structure-activity relationships for biologically active congeneric compounds were examined by using free-energy-related physicochemical parameters and regression technique. The steric effects involved in such biological activities were expressible and separable from other factors as steric parameters depending upon the situations. In this article, examples for enzyme reactions, enzyme inhibitions, insecticides, fungicides, herbicides, plant growth regulators and synthetic sweetners, mostly from our own laboratory are reviewed. The steric parameters used here are Taft E_s, Hancock E_s^c, Verloop STERIMOL and van der Waals molar volume.

Abbreviations

AChE	acetylcholinesterase
DDD	p,p'-dichlorodiphenyldichloroethane
DDT	p,p'-dichlorodiphenyltrichloroethane
HB	hydrogen bonding parameter
MAO	monoamine oxidase
MBC	minimum blocking concentration
MEC	minimum excitatory concentration
QSAR	quantitative structure-activity relationship

1 Introduction

In recent years, quantitative procedures have been developed for the structure-activity correlation studies of biologically active compounds. Among them, the Hansch approach has been most widely and effectively used and sometimes successfully applied to design novel compounds of potent activity.

When the Hansch approach was initiated, the correlation of the biological activity (1/C; C is the equieffective concentration or dose) with structural parameters was analyzed using Eq. 1,

$$\log(1/C) = a\pi - b\pi^2 + \varrho\sigma + \text{constant} \quad (1)$$

where only electronic (σ) and hydrophobic (π) factors were considered as determining the activity [1]. However, the importance of the steric factor in biological activity suggested by the classical "lock and key" theory was never forgot. The steric factor was first successfully parameterized by Hansch, Deutsch and Smith [2] in Eq. 2 with the Taft E_s value for substituent effects on enzymic reactions.

$$\log(1/C) = a\pi + \varrho\sigma + \delta E_s + \text{constant} \quad (2)$$

In the following years, various steric parameters have been applied to the analysis besides the Taft E_s value. For instance, the Hancock "corrected" steric E_s^c, the Bondi molecular volume V_w, and the Verloop STERIMOL parameters have been used to rationalize various steric effects depending upon the interactions involved. At every addition of such parameters, the versatility of the approach has been expanded remarkably. The number of successful applications has been growing enormously leading to the present state of development.

For years, we have been studying QSAR (quantitative structure-activity relationship) analyses of pesticides and other bioactive compounds. In many examples, we have found a decisive role of the steric effect in determining the activity variation. In this chapter, applications of various steric constants such as E_s, E_s^c, V_w and STERIMOL parameters to QSAR studies mostly from our own laboratory are reviewed.

The molecular refractivity, MR, has been successfully utilized in QSAR studies [3]. This parameter is, in a way, an index for the molecular volume. However, since it is originally defined as a measure of the polarizability related to dispersion forces, it is not included.

2 The Taft E_s Constant

2.1 The Nature of the E_s Constant

The variation in the acid-catalyzed hydrolysis of aliphatic carboxylic esters RCOOEt is mostly subject to the steric effect intramolecularly exerted by the substituent R on the formation of the tetrahedral reaction intermediate. Taft [4] defined the steric parameter E_s as Eq. 3,

$$E_s = \log(k_R/k_{Me})_A \quad (3)$$

where $(k_R/k_{Me})_A$ refers to the ratio of acid catalyzed hydrolytic rate constants of RCOOR′ to that of MeCOOR′. The bulkier the substituent, the more negative the E_s value. By definition, $E_s(Me) = 0$. He recognized that the E_s parameter varies in parallel with the group radius. Charton examined the relationship between them quantitatively [5], and showed that the E_s value for each set of substituents of the types, CH_2X, CHX_2 and CX_3 including X = H, is linearly dependent on van der Waals radius, $r_v(X)$ of the hetero atom X being halogens or O and S in OMe and SMe groups.

Kutter and Hansch elaborated the relationships and developed Eq. 4 for symmetric top-type groups such as CX_3 and H [6].

$$E_s = \underset{(0.55)}{3.483} - \underset{(0.22)}{1.839} r_v \qquad n = 6 \qquad s = 0.132 \qquad r = 0.996 \qquad (4)^1$$

For CX_3, r_v was taken as the average of the maximum and minimum van der Waals radii from the axis. The E_s values of the hetero atom substituents, which cannot be obtained by Taft's original method, were estimated using Eq. 4 not only for other symmetric-top monoatomic substituents as halogens but also for unsymmetric substituents like NR_2, OR, SR, NO_2, Ac and Ph. For NR_2, OR, and SR groups, the values were calculated from the N, O and S radius only. For NO_2, Ac and Ph, two values of E_s were estimated corresponding to the group being either coplanar with or perpendicular to the reaction site by substituting the half width or thickness of the groups for r_v in Eq. 4 [6].

These E_s parameters estimated by Eq. 4 for hetero atom substituents can be combined with those originally developed for various alkyl groups as a set of steric constants for QSAR studies of aromatic systems [6]. Thus, apart from the original definition for the intramolecular steric effect, the combined set of E_s parameters is able to represent intermolecular steric effects as well. The original Taft E_s values for unsymmetrical alkyl groups seem to represent effective steric dimension of the groups which is scaled on the same standard as those for symmetric monoatomic substituents where the "effective" dimension coincides with the van der Waals radius.

Taft also defined the E_s parameter for aromatic ortho substituents in a fashion similar to Eq. 3 [4]. Charton advanced evidence that the original value E_s^{ortho} is linked rather with the polar substituent effect despite the original definition [5]. We have shown that the set of E_s parameters defined by Eq. 3 and 4 can also be used to rationalize the steric effect of aromatic ortho substituents [7]. In general, the ortho substituent effect in biological activity is analyzable as linear combination of substituent parameters [8] (Eq. 5),

$$\log(1/C) = a\pi - b\pi^2 + \varrho\sigma + \delta E_s + fF + \text{constant} \qquad (5)$$

although each of the terms is not always statistically significant. In Eq. 5, σ is taken as σ_{para}. F is a parameter for inductive effect defined by Swain and Lupton [9] and corrected by Hansch et al. [10]. It is supposed to take care of proximity electronic effect of ortho substituents. The term E_s in Eq. 5 represents not only intramolecular proximity types but also intermolecular steric effects. Depending on the steric effect, however, E_s can be replaced by other parameters.

[1] In Eq. 4 and the following equations, n is the number of compounds used in the regression, s is the standard deviation and r is the correlation coefficient. The figures in the parentheses are the 95% confidence intervals.

2.2 Applications of the E_s Constant

2.2.1 Monoamine Oxidase Inhibitors

The QSAR analysis of the monoamine oxidase (MAO) inhibitors by Kutter and Hansch is one of the earliest successes in the application of E_s constant [6]. They

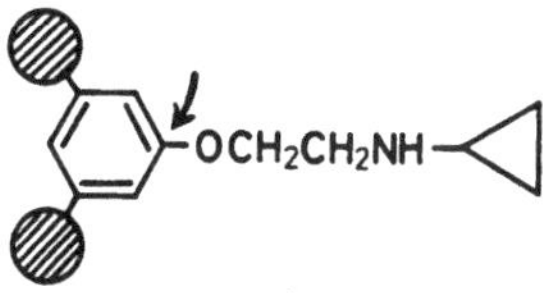

1

developed Eq. 6 for inhibitory activity of N-(substituted phenoxyethyl)cyclopropylamines (*1*) against MAO prepared from rat liver.

$$pI_{50}(M) = \underset{(0.15)}{0.766\,\Sigma E_s^{3,5}} + \underset{(0.40)}{1.752\,\Sigma\sigma} + \underset{(0.18)}{0.180\,\Sigma\pi} + \underset{(0.30)}{3.996} \quad (6)$$
$$n = 15 \quad s = 0.203 \quad r = 0.976$$

While $\Sigma\sigma$ and $\Sigma\pi$ are the sum of values for substituents irrespective of their positions, $\Sigma E_s^{3,5}$ is the sum only for 3- and 5-substituents. Eq. 6 shows that the bulkier the meta substituents, the lower the MAO inhibition activity. Hydrophobic as well as electron withdrawing substituents favor the activity.

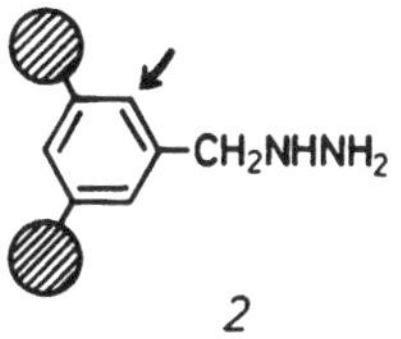

2

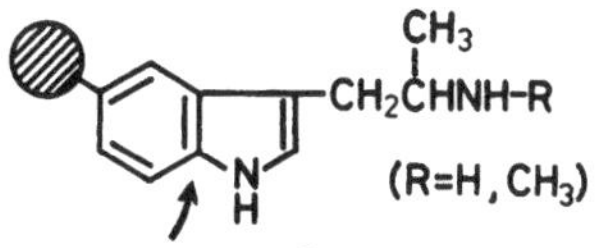

3

We also studied various series of MAO inhibitors [11] and formulated Eq. 7 and 8 for the data of substituted benzylhydrazines (*2*) and α-methyltryptamines (*3*), respectively, against MAO prepared from guinea pig liver mitochondria.

$$pI_{50} = \underset{(0.125)}{-0.545\,\Sigma\pi} + \underset{(0.271)}{0.638\,\Sigma\sigma_2} + \underset{(0.161)}{0.516\,\Sigma E_s^{3,5}} + \underset{(0.209)}{5.832} \quad (7)$$
$$n = 8 \quad s = 0.06 \quad r = 0.996$$

$$pI_{50} = \underset{(0.620)}{-1.085\,\Sigma\pi_{4,6}} + \underset{(0.714)}{1.251\,\Sigma\sigma_{7a}} + \underset{(0.438)}{1.071\,E_s^5} + \underset{(0.400)}{3.152} \quad (8)$$
$$n = 15 \quad s = 0.23 \quad r = 0.862$$

In Eq. 7 and 8, $\Sigma\sigma_2$ and $\Sigma\sigma_{7a}$ are the sum of σ constant of substituents directed to the ortho and the 7a position, respectively. In Eq. 7, $\Sigma\pi$ is the sum for all substituents, whereas, in Eq. 8, $\Sigma\pi_{4,6}$ is the sum for 4- and 6-substituents. The comparison of Eq. 7 and 8 indicates that the aromatic substituents of benzylhydrazines and α-methyltryptamines exert similar hydrophobic, electronic and steric effects on guinea pig liver mitochondrial enzyme.

Except for the $\Sigma\pi$ term indicating that the hydrophobicity of substituents is unfavorable to the guinea pig enzyme inhibition, Eq. 7 and 8 are also very similar to Eq. 6 for the rat liver mitochondrial enzyme inhibition of phenoxyethylcyclopropylamines. In the structural formulae *1*, *2* and *3*, arrows indicate the position to which the electronic effect is directed and shaded circles show the substituents exerting a steric effect. The position-specific steric effect of aromatic substituents is similar among three sets of inhibitors. The E_s parameters used for unsymmetrical groups such as NO_2, OR and NR_1R_2 are either derived from their half thickness or from the van der Waals radius of just oxygen and nitrogen. Thus, the steric effect would be due to fit a surface rather to engulf substituents. Very similar substituent effects were also observed in other sets of various MAO inhibitors [11].

2.2.2 Enzyme Reactions

There are quite a few enzyme reactions of aromatic substrates and inhibitors where the steric effect of aromatic substituents is separated from others with E_s constant. Eq. 9 was developed for the Michaelis constant, K_m, of substituted D-phenylglycines with hog kidney D-phenylglycine oxidase [11].

$$\begin{array}{l} \log(1/K_m) = 0.300\pi_3 + 0.593\sigma + 0.212E_s^3 + 2.339 \\ \qquad\qquad\quad (0.236) \quad (0.321) \quad (0.181) \quad (0.164) \\ n = 14 \quad s = 0.161 \quad r = 0.860 \end{array} \tag{9}$$

The terms π_3 and E_s^3 indicate hydrophobic and steric effects specific to m-substituents. Eq. 10 describes substituent effects on the rate of transfer of substituted anilines catalyzed by carp vicera thiaminase [11].

$$\begin{array}{l} \log k = 0.722\pi - 1.842\sigma^- + 0.557E_s^4 - 1.892 \\ \qquad\quad (0.377) \quad (0.484) \quad (0.404) \quad (0.379) \\ n = 12 \quad s = 0.288 \quad r = 0.959 \end{array} \tag{10}$$

The reaction schemes for Eq. 9 and 10 are shown in Eq. 11 and 12, respectively.

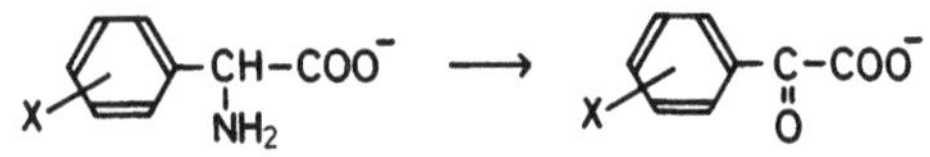

(11)

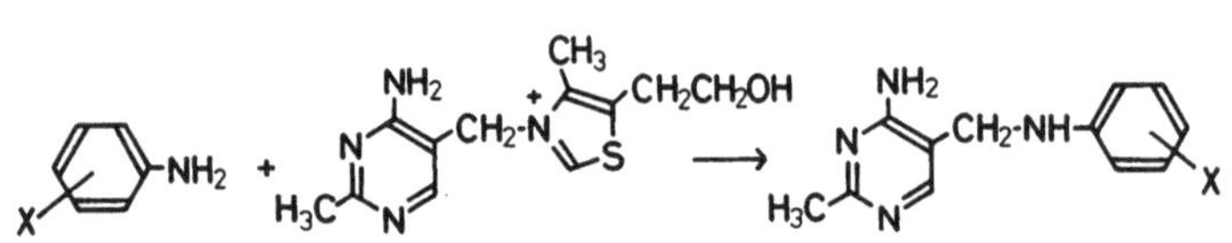

(12)

The substrate specificities for these enzyme reactions were originally based only on electronic effects of substituents [12,13]; the chemical mechanism of action depending on the electron-withdrawing ability of aromatic substituents has been postulated as being expressible by biphasic Hammett plots such as Fig. 1 for the reactions of Eq. 12. Eq. 9 and 10 clearly indicate the significance not only of electronic but also of steric and hydrophobic effects.

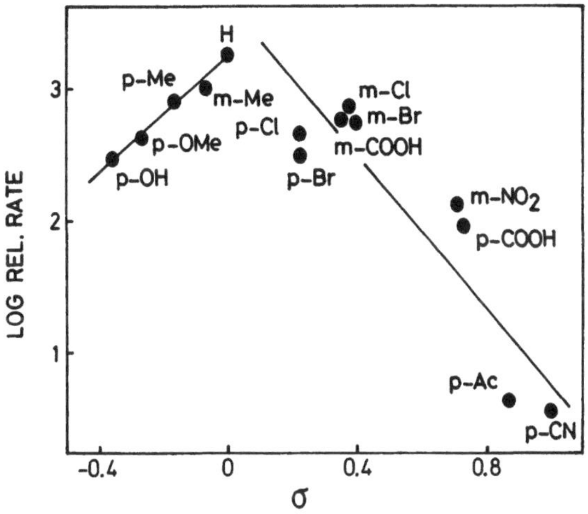

Fig. 1. Apparent biphasic Hammett plot for the rate of transfer of substituted anilines catalyzed by carp vicera thiaminase (reproduced from Mazrimas et al. [13] with permission from the authors and Academic Press, Inc.)

2.2.3 Phenyl N-Methylcarbamates as Acetylcholinesterase Inhibitors

Substituted phenyl N-methylcarbamates (ArOCONHMe) are one of the most widely used agricultural insecticide classes which inhibit acetylcholinesterase (AChE). The steps by which they react with AChE are shown in Eq. 13, where EOH denotes the enzyme.

$$\text{ArOCONHMe} + \text{EOH} \underset{k_{-1}}{\overset{k_1}{\rightleftharpoons}} \begin{matrix}\text{Reversible}\\ \text{complex}\end{matrix} \xrightarrow{k_2} \begin{matrix}\text{EOCONHMe}\\ + \\ \text{ArOH}\end{matrix} \qquad (13)$$

We have studied the molecular mechanism of the enzyme inhibition using bovine erythrocyte AChE [14]. Depending upon the position and nature of substituents, the value of k_1/k_{-1} ($= 1/K_d$) showed significant variations whereas that of k_2 did not. Thus, formation of the reversible complex was considered to be the step which governs the variation in overall inhibitory activity.

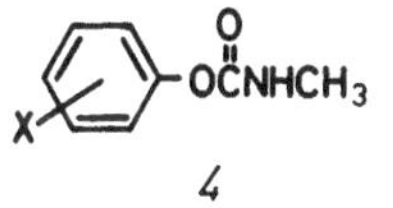

4

Eq. 14 is the result of analysis for 53 compounds (*4*) having such diverse substituents as alkyl, branched alkyl, alkoxy, branched alkoxy, acyl, NO_2, CN, SR, SO_2R, halogen and CF_3 at o-, m- and p-positions.

$$\begin{aligned}\log(1/K_d) = {} & \underset{(0.168)}{1.399\pi_{2,3}} + \underset{(0.159)}{0.306\pi_4} + \underset{(0.380)}{1.659\sigma^{o}_{\varrho>0}} - \underset{(0.371)}{1.784\sigma^{o}_{\varrho<0}} \\ & + \underset{(0.132)}{0.168E_s^{ortho}} + \underset{(0.536)}{0.770F_{ortho}} + \underset{(0.248)}{1.358HB} + \underset{(0.162)}{2.592} \qquad (14)\end{aligned}$$

$$n = 53 \quad s = 0.238 \quad r = 0.947$$

The subscripts 2, 3 and 4 indicate substituents at the o-, m- and p-positions, respectively. The slopes of $\pi_{2,3}$ and π_4 terms suggest that the hydrophobic nature of the enzyme surface corresponding to o- and m-positions is approximately equivalent to each other, and higher than that of the surface corresponding to the p-position.

σ^o is the electronic constant which supposedly contains no through-resonance effect. The substituents are classified in terms of electronic effect into two groups: those in one group are more electron withdrawing and promote an attack by a nucleophile of the enzyme on the carbonyl carbon of the carbamyl group, and those in the other group are more electron-releasing and assist an electrophilic attack by an acidic group of the enzyme on the carbonyl oxygen atom. Substituents in the first group are those at the o-position, and those which are electron-withdrawing at the m- and p-positions, such as NO_2, CN and acyl. Their electronic effect is expressed by the $\sigma^o_{\varrho>0}$ term. All other substituents belong to the second group, the electronic effect of which is represented by the $\sigma^o_{\varrho<0}$ term. The significance of these two terms in Eq. 14 suggests different mechanisms for the two groups of substituents, leading to a common tetrahedral intermediate as shown in Fig. 2. Electron-donating o-substituents do not follow the negative ϱ mechanism because the acid-catalytic site of the enzyme does not fit the carbonyl oxygen atom due to hindrance exerted by these o-substituents.

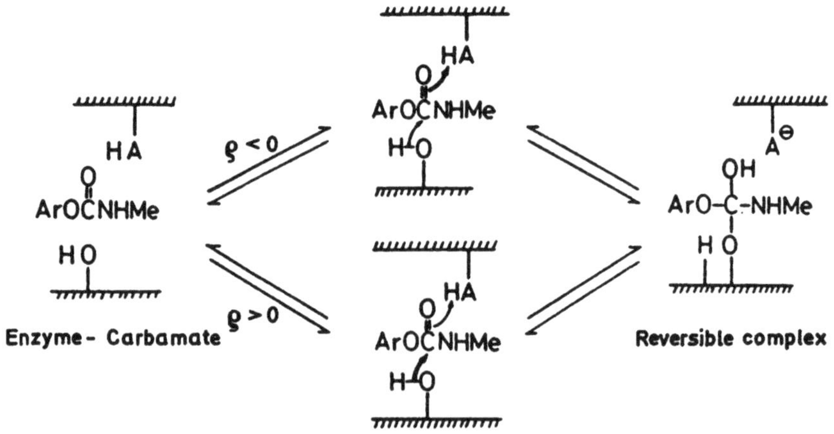

Fig. 2. Two reaction pathways of phenyl N-methylcarbamates with acetylcholinesterase; the *bold-faced arrow* indicates an electron pair migration as the driving force of the reaction [14] (reproduced with permission from Academic Press, Inc.)

In Eq. 14, the reference of E_s is shifted to that of H. The proximity effects of ortho substituents are well separated by E_s and F terms according to Eq. 5. The coefficient values of E_s and F terms, 0.17 and 0.77, are close enough to those for the alkaline hydrolysis of o-substituted phenyl acetates [15]. Thus, in support of the above discussion, the proximity effects of o-substituents are considered to be those on formation of the tetrahedral intermediate.

HB, an indicator variable for hydrogen bonding effect of substituents [16], is 1 for hydrogen bonding substituents such as o-OR, m-acyl, -CN, $-NO_2$ and $-NMe_2$, but otherwise is zero. The significance of the term in Eq. 14 indicates a specific hydrogen bond formation of these groups with a hydrogen donor on the enzyme. The hydrogen donor site is supposed to be located unsuitably for interaction with other hydrogen bonding groups such as $o\text{-}NO_2$, CN and m-OR. Fig. 3 shows the stereospecific situation schematically.

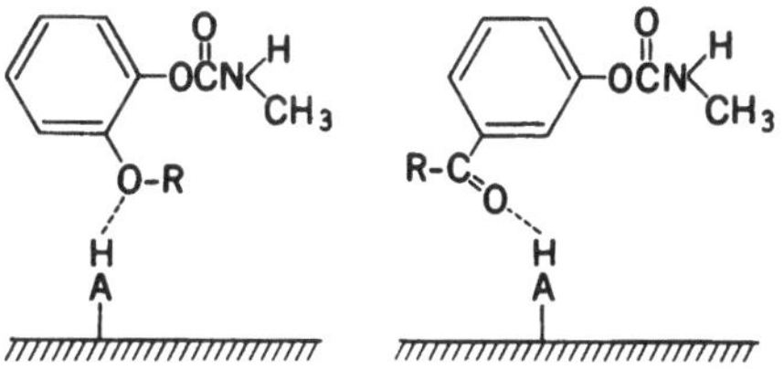

Fig. 3. Stereospecific hydrogen bond formation of phenyl N-methylcarbamates with acetylcholinesterase [14] (reproduced with permission from Academic Press, Inc.)

2.2.4 Herbicidal N-Chloroacetyl-N-phenylglycine Esters

Chloroacetamide derivatives such as N,N-diallyl (*5*) and N-alkoxymethyl-N-2,6-diethylphenyl (*6*) analogs are widely used as upland field herbicides. By modification of chloroacetamide structure, Fujinami et al. found that the N-chloroacetyl-N-phenylglycine esters (*7*) show varying degrees of inhibitory activity, in particular, against shoot growth of annual grasses [17]. They developed Eq. 15 and 16 for the herbicidal activities of various esters where the aromatic substituents are fixed

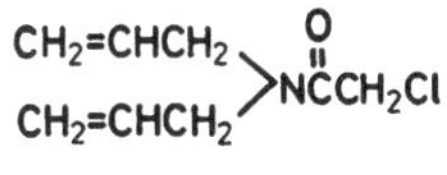

5

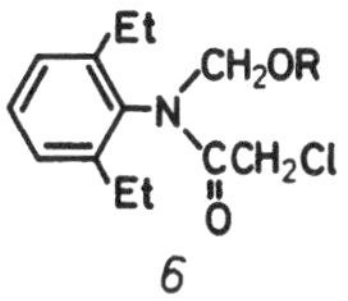

6

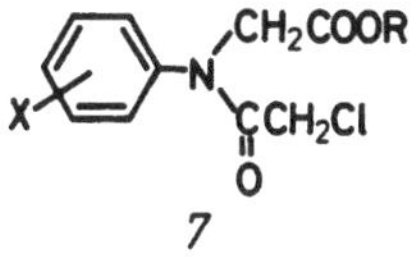

7

as 2,6-diethyl against the rice plant, *Oryzae sativa var. Waseasahi*, and barnyard grass, *Echinochloa cruss-galli var. frumentaceus*, respectively. The ester substituents include not only simple alkyl and branched alkyl but also substituted benzyl, alkoxyalkyl, cyanoalkyl and carbamoyalkyl.
The rice plant:

$$\begin{aligned} pI_{50} &= -0.278\,(\log P)^2 + 1.474 \log P - 1.949\sigma^* + 4.000 \\ &\quad\ (0.163) \qquad\quad (1.131) \qquad (0.572) \qquad (1.902) \\ n &= 22 \qquad r = 0.922 \qquad s = 0.359 \end{aligned} \tag{15}$$

Barnyard grass:

$$\begin{aligned} pI_{50} &= -0.290\,(\log P)^2 + 1.884 \log P - 0.996\sigma^* + 2.885 \\ &\quad\ (0.170) \qquad\quad (1.185) \qquad (0.600) \qquad (1.993) \\ n &= 22 \qquad r = 0.850 \qquad s = 0.376 \end{aligned} \tag{16}$$

In these equations, I_{50} is the molar concentration which inhibits the shoot elongation of the plants to half the length of the control. The activites vary parabolically with the change of log P value. The optimum log P value is 2.6 for the rice plant and 3.3 for barnyard grass. This situation is schematically shown by Fig. 4, illustrating that the selectivity in terms of ΔpI_{50} increases with increasing log P value. Thus, there is a range with sufficient herbicidal activity against barnyard grass within which compounds with the lowest possible toxicity against the rice plant can be designed. Fujinami et al.[17] suggested compounds having an ester moiety of C_3—C_5 as possible candidates.

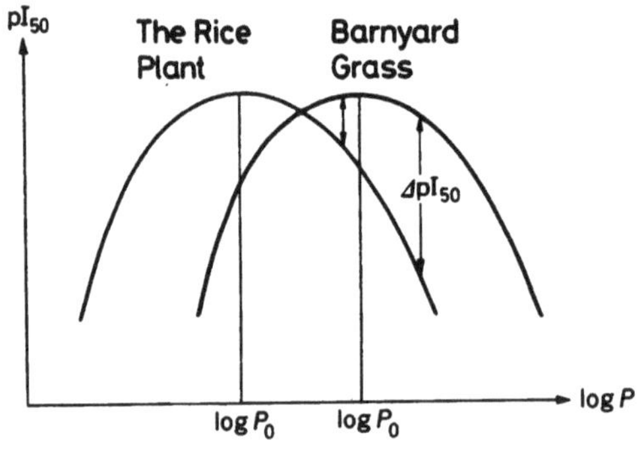

Fig. 4. Parabolic relationships between herbicidal activity and hydrophobicity and the selectivity between two plant species

Similar analysis of compounds having a fixed ethyl ester moiety and variated aromatic substituents gave Eq. 17 and 18.
The rice plant:

$$\begin{aligned} pI_{50} &= -0.328 \log P - 0.950 E_s^{\circ} - 0.618 \Sigma\sigma + 4.625 \\ &\quad\ (0.182) \qquad\quad (0.142) \qquad (0.458) \qquad (0.193) \\ n &= 28 \qquad r = 0.959 \qquad s = 0.261 \end{aligned} \tag{17}$$

Barnyard grass:

$$pI_{50} = -0.767\,E_s^o - 0.218\,E_s^m + 3.990 \quad (18)$$
$$\qquad\quad (0.158) \qquad (0.196) \qquad (0.241)$$
$$n = 28 \quad r = 0.905 \quad s = 0.295$$

The negative coefficient of E_s^o and E_s^m indicates an enhancement of activity with an increase of steric bulk. Especially, the variation in activity against barnyard grass is mainly governed by steric factors. If steric bulkiness prevents the proper fit with the target site, these results mean that it inhibits the interaction with some enzymes which participate in the detoxication process(es) in the plant body. The negative slope of the σ^* term in Eq. 15 and 16 coincides with this view, i.e. a molecule with a more electron-withdrawing substituent becomes more susceptible to a hydrolytic attack.

2.2.5 Fungicidal Cyclic N-Phenylimides

In cyclic N-phenylimides (*8* and *9*), quite a few derivatives such as procymidon (*8*: X = 3,5-Cl_2) and vinclozolin (*9*: X = 3,5-Cl_2) have been found to be active as

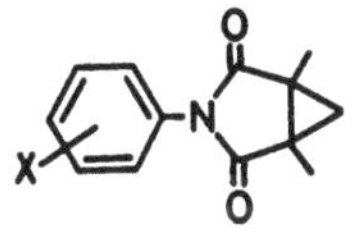

8

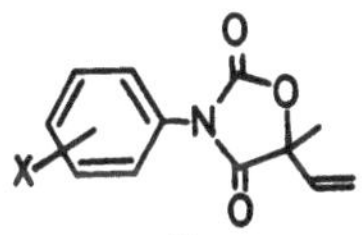

9

potent agricultural fungicides of practical use. Takayama et al.[18] examined the aromatic substituent effect of these classes of fungicides. They showed that antifungal activity against *Botrytis cinerea*, in terms of pI_{50} value, corresponds very nicely between N-phenyl-1,2-dimethylcyclopropanedicarboximides (*8*) and N-phenyl succinimides (*10*) with the same aromatic substituents, indicating identical substituent

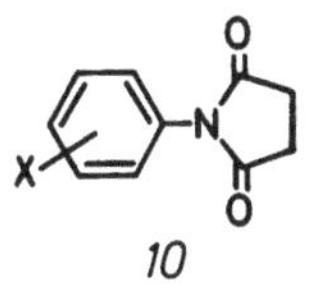

10

effects between the two series. The structure-activity correlation for the N-phenyl succinimides is shown in Eq. 19,

$$pI_{50} = 0.723\,\Sigma\pi_{3,5} + 1.464\,\Sigma\sigma^o + 0.894\,\Sigma E_s^{2,6} + 0.671\,E_s^m$$
$$\qquad (0.201) \qquad (0.266) \qquad (0.195) \qquad (0.235)$$
$$+ 0.345\,E_s^p - 0.543\,HB + 3.690 \quad (19)$$
$$\qquad (0.207) \qquad (0.183) \qquad (0.233)$$
$$n = 61 \quad r = 0.952 \quad s = 0.293$$

where the $\Sigma\pi_{3,5}$ and $\Sigma E_s^{2,6}$ are the sums of π and E_s for substituents at designated positions and $\Sigma\sigma^o$ is the sum of the σ^o values of all ring substituents. The E_s^m is that of either 3- or 5-substituents having bulkier dimensions in terms of E_s. HB is the indicator variable for hydrogen bonding substituents such as OR, COR, COOMe, NO_2 and CN [16]. Electron-withdrawing effects favor fungicidal activity, but the hydrogen-bonding factor reduces it. The hydrophobic effect is position-specific. Only those of the m-substituents are significant and they additively enhance the activity. The steric effect is also position-specific. The positive signs of E_s indicate that the bulkier the substituent at any position, the lower the activity, other factors being equal.

The high antifungal activity exerted by N-(3,5-dihalophenyl) derivatives was nicely rationalized by the hydrophobic and electron-withdrawing properties of halogen atoms, and also because the steric bulk of only one of the two halogen substituents is unfavorable to activity. Substantially the same effects are thought to be at work in the interaction of the N-phenyl dimethylcyclopropanedicarboximides (*8*) with the critical site, and it is also expected to be so with the other N-phenylimide-types of fungicides, such as 3-phenyloxazolidine-2,4-diones (*9*).

3 The "Corrected" Steric E_s^c Constant

3.1 The Nature of the E_s^c Constant

Since the E_s value is determined by the relative activation free energy from the unsaturated initial state to the saturated tetrahedral intermediate state of the ester hydrolysis, Hancock and his coworkers considered that a hyperconjugation effect of α-hydrogen may contribute to the estimate of E_s values [19]. To separate the hyperconjugation effect from the "true steric effect", they defined the parameter E_s^c (corrected steric) as Eq. 20, assuming that the hyperconjugation effect is proportional to the number of α-hydrogen atoms, n_H. By definition, $E_s^c(\text{Me}) = 0$.

$$E_s^c = E_s + 0.306\,(n_H - 3) \qquad (20)$$

We examined whether the steric effect of substituents R of type $CR^1R^2R^3$ can be expressed in terms of the steric effect of component substituents R^1, R^2 and R^3 using E_s^c parameters. With increasing substitution at the α-carbon, the total steric effect of substituents R has been observed to increase telescopically in such series as Me, Et, i-Pr and t-Bu [4]. Thus, the simple addition of steric parameters for α-substituents is inadequate to represent the situation. We have found that the E_s^c parameter for 24 primary, secondary and tertiary alkyl groups can be formulated as Eq. 21 by a linear combination of E_s^c parameters of component α-substituents [20].

$$\begin{aligned} E_s^c(CR^1R^2R^3) = {} & \underset{(0.516)}{3.429\,E_s^c(R^1)} + \underset{(0.252)}{1.978\,E_s^c(R^2)} + \underset{(0.118)}{0.649\,E_s^c(R^3)} \\ & \underset{(0.195)}{-\,2.104} \end{aligned} \qquad (21)$$

$$n = 24 \quad s = 0.191 \quad r = 0.992$$

The substituents R^1, R^2 and R^3 are classified according to the relative magnitude of their E_s^c value so that $E_s^c(R^1) \geqq E_s^c(R^2) \geqq E_s^c(R^3)$, i.e., R^1 is the smallest while R^3 is the bulkiest. The E_s^c value of the α-hydrogen substituent of primary and secondary alkyl groups is taken as $E_s(H) - 3 \times 0.306 = 0.32$. In deriving Eq. 21, a few component substituents were considered as being conformationally restricted. Their effective steric effect was represented not by their original E_s^c but the values for substituents whose geometry is similar to the restricted conformation. The analysis with the use of E_s instead of E_s^c in Eq. 21 was shown to give a poorer result.

Recently, Dubois et al. reexamined this type of correlations [21]. They showed that E_s and E_s^c values of $R^1R^2R^3C$ are analyzable equally well with those of component groups by Eq. 22 and 23, respectively, where the contribution of the Δn ($= n_H - 3$) term for each component group is not fixed as in Eq. 20 but made adjustable.

$$E_s(R^1R^2R^3C) = \Sigma[a_iE_s(R_i) + b_i\Delta n_i] + a_o \qquad (22)$$

$$E_s^c(R^1R^2R^3C) = \Sigma[a_i'E_s^c(R_i) + b_i'\,\Delta n_i] + a_o' \qquad (23)$$

In the preceding section, the Taft parameter E_s was shown to be suitable for the type of steric effects due to "effective" van der Waals width of substituents without the correction for the "hyperconjugation" effect. Since the "uncorrected" E_s value has its own significance as a steric parameter, the "correction" term, $0.306\,(n_H - 3)$, should be interpreted in another way. Proportional to the number of α-branches of substituents, the E_s^c is more negative than the corresponding E_s by a factor 0.306. Thus, it emphasizes the effect of α-branching of alkyl groups more than E_s. For the E_s^c value of hydrogen, n_H is taken as zero so that $E_s^c(H) = 0.32$ while $E_s(H) = 1.24$. Thus, the E_s^c scale estimates the effective bulk of Me relative to that of H as smaller than does the E_s scale. We postulate that the more strict the steric demands between interacting partners, the more suitable the use of E_s^c scale than the use of E_s.

Although a fixed value, 0.306, is not necessarily adequate and the best factor had better to be determined by the regression analysis using Eq. 22 or its counterparts, we used E_s^c values defined by Eq. 20 for the steric effect in the following examples. The E_s^c value works very well for a number of examples. This might be simply fortuitous but rather favorable in reducing the number of independent variables for the analysis.

Since the total steric effect of component substituents R^1, R^2 and R^3 on the reaction center in the transition state of ester hydrolysis is similar, if not identical, to that of three N-substituents of aliphatic amines on a certain electron acceptor or electrophile, we attempted to separate electronic and steric effects of N-substituents of amines $NR^1R^2R^3$ for various reactions by Eq. 24 where K is either the rate or equilibrium constant, $\Sigma\sigma^*$ is the sum of σ^* values for R^1, R^2 and R^3 and $E_s^c(R^1) \geqq E_s^c(R^2) \geqq E_s^c(R^3)$, and showed that Eq. 24 is generally applicable [22].

$$\log K = \varrho^* \Sigma\sigma^* + \delta_1E_s^c(R^1) + \delta_2E_s^c(R^2) + \delta_3E_s^c(R^3) + \text{constant} \qquad (24)$$

Examples are given in Eq. 25 and 26.

Hydrogen bond formation with $CHCl_3$ in cyc-hexane at 35 °C:

$$\begin{aligned}\log K = &-0.499\,\Sigma\sigma^* + 1.013\,E_s^c(R^1) + 0.419\,E_s^c(R^2)\\ &\quad(0.210)\qquad\quad(0.229)\qquad\qquad(0.108)\\ &+ 0.209\,E_s^c(R^3) - 0.101\\ &\quad(0.098)\qquad\qquad(0.138)\\ n = 29 \qquad & s = 0.145 \qquad r = 0.946\end{aligned} \tag{25}$$

Association with Me_3B in gas phase at 100 °C:

$$\begin{aligned}\log K = &-4.878\,\Sigma\sigma^* + 14.585\,E_s^c(R^1) + 4.879\,E_s^c(R^2)\\ &\quad(2.087)\qquad\quad(5.655)\qquad\qquad(2.578)\\ &+ 1.461\,E_s^c(R^3) + 0.001\\ &\quad(0.594)\qquad\qquad(0.755)\\ n = 17 \qquad & s = 0.569 \qquad r = 0.876\end{aligned} \tag{26}$$

In general, $\delta_1 \geqq \delta_2 \geqq \delta_3$ indicating that the steric effect of the smallest component is most significant in determining the total steric effect.

We have determined the ion-pair formation-partition equilibrium constant with picrate anion for a number of primary, secondary, tertiary and quaternary ammonium ions [23]. In aqueous media of pH 5–6, the ammonium ions and picrate are considered to exist almost completely as unpaired counter ions. When the aqueous solution is mixed with an immiscible organic solvent, the ions are partitioned into the organic phase as the ion pair. We expected that the steric effect of N-substituents in the ion-pair formation-partition equilibrium could be analyzed by a procedure similar to Eq. 24, and derived Eq. 27 for the set of quaternary ions [23].

$$\begin{aligned}\log K = &\ 0.870\,\Sigma\pi + 0.848\,E_s^c(R^1) + 0.677\,E_s^c(R^2) + 0.225\,E_s^c(R^3)\\ &\ (0.088)\qquad(0.287)\qquad\qquad(0.244)\qquad\qquad(0.166)\\ &- 1.774\\ &\quad(0.269)\\ n = 31 \qquad & s = 0.109 \qquad r = 0.990\end{aligned} \tag{27}$$

In Eq. 27, K is the equilibrium constant (M^{-1}). The four N-substituents are classified as $E_s^c(R^1) \geqq E_s^c(R^2) \geqq E_s^c(R^3) \geqq E_s^c(R^4)$. Hydrophobic substituent parameter, π, is evaluated by taking the H substituent as the reference, and summed for component four substituents. Since the steric effect of bulkiest R^4 substituents is not significant, the ion-pairing is perhaps achieved in such a manner that the counter anion must approach from the least hindered side of the ammonium ions (*11*).

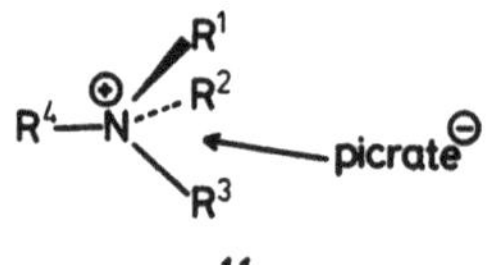

11

Eq. 28 was derived from the combined set of primary, secondary, tertiary and quaternary ammonium ions.

$$\begin{aligned}\log K = \ & \underset{(0.059)}{0.899\Sigma\pi} + \underset{(0.201)}{1.049E_s^c(R^1)} + \underset{(0.157)}{0.682E_s^c(R^2)} + \underset{(0.107)}{0.235E_s^c(R^3)} \\ & + \underset{(0.041)}{0.495n_H} - \underset{(0.187)}{1.852} \qquad (28)\\ n = 58 \quad & s = 0.114 \quad r = 0.989\end{aligned}$$

In this equation, n_H is the number of N-hydrogen substituents. For the set of primary, secondary and tertiary ions, R^1 is always H. The steric effect of H as the R^1 substituent is delineated by the $E_s^c(R^1)$ term in Eq. 28 along with that of R^1 alkyl substituents in quaternary ammonium ions. The hydrogen bonding association for primary, secondary and tertiary ions of the type $\equiv N^{\oplus}—H \cdots O^{\ominus}C_6H_2(NO_2)_3$ is not likely to occur in solvents. Thus, the n_H term may indicate that the greater the number of NH hydrogen, the more stable the hydrogen bonding with the more basic 1-octanol favoring the ion-pair partitioning into the organic phase. The coulombic force between counter charges is the driving factor for the ion pair formation. However, no electronic term is discernible in Eq. 28 showing that the positive charge distribution does not change significantly with the structural variation as far as alkyl ammonium ions are concerned.

3.2 Applications of the E_s^c Constant

3.2.1 Earlier Examples of the Application

The E_s^c parameter was first introduced by Hansch and Lien into biological structure-activity studies [24]. They showed that the adrenergic blocking activity of β-haloalkyl-

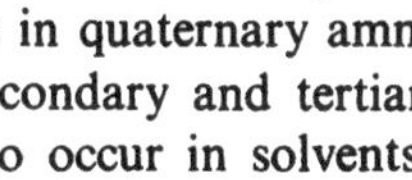

12

amines (*12*) are well analyzed in terms of parameters for N-substituents as Eq. 29.

$$\begin{aligned}-\log ED_{50}\,(\text{mole/kg cat}) = \ & \underset{(1.82)}{3.57\Sigma\sigma^*} + \underset{(0.43)}{1.11\Sigma E_s^c} \\ & - \underset{(1.09)}{4.43n_H} + \underset{(1.40)}{11.91} \qquad (29)\\ n = 10 \quad s = 0.235 \quad & r = 0.986\end{aligned}$$

Other earlier analyses made by Hansch et al. are represented by Eq. 30 and 31.

Hydrolysis of p-nitrophenyl alkanoates (*13*) with serum esterase [25]:

$$\begin{aligned}\log(\text{Relative Rate}) = \ & \underset{(7.20)}{-15.65\sigma^*} + \underset{(0.56)}{2.76E_s^c} + \underset{(0.67)}{1.77} \qquad (30)\\ n = 6 \quad s = 0.232 \quad & r = 0.995\end{aligned}$$

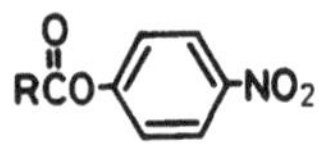

13

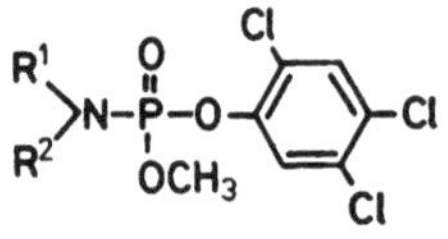

14

Inhibition of fly-head acetylcholinesterase by phosphoramidates (*14*)[25]:

$$\log K_i = 1.08\Sigma\sigma^* + 1.00\Sigma E_s^c + 5.46$$
$$\quad (1.30) \quad (0.42) \quad (0.64) \tag{31}$$
$$n = 8 \quad s = 0.309 \quad r = 0.970$$

One example performed in this laboratory is the analysis of toxicity exerted by bicyclic phosphate esters[26]. 2,6,7-Trioxa-1-phosphabicyclo[2.2.2]octane-1-oxides (*15*) with suitable 4-substituents (R) are highly toxic to mammals[27]. The 4-Et derivative

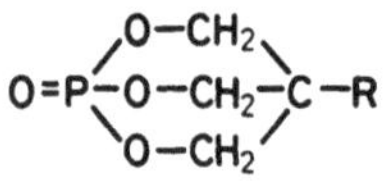

15

is the toxic principle in the smoke produced on burning a noncommercial fire-retardant polyuretan foam[28]. The analysis of toxicity in terms of LD_{50} (mole/kg mice) determined after 24 hrs of injection gave Eq. 32.

$$\log (1/LD_{50}) = -1.25\pi^2 + 3.51\pi + 0.49\sigma^* - 0.68E_s^c + 2.65$$
$$\quad (0.25) \quad (0.91) \quad (0.45) \quad (0.18) \quad (0.74) \tag{32}$$
$$n = 18 \quad s = 0.246 \quad r = 0.976$$

The sign of the E_s^c term is negative indicating that the bulkier the substituent, the higher is the perturbation with the site(s) of action leading to the higher toxicity. This class of compounds were recently suggested to antagonize the actions of synaptically released GABA[29].

3.2.2 Applications to Development of Pesticides

Recently, Kirino and coworkers analyzed the activity of N-alkylaminoacetonitriles (*16*) to prevent "yellows" of the Japanese radish, a disease caused by a soil-borne

$RNHCH_2CN$

16

fungus, *Fusarium oxysporum*, and derived Eq. 33, where ED_{50} is the 50% effective dose (mole) in preventing the disease [30].

$$pED_{50} = \underset{(0.184)}{-0.606 E_s^c} + \underset{(0.265)}{1.518} \qquad (33)$$
$$n = 16 \quad s = 0.204 \quad r = 0.884$$

Since the activity determination requires a few weeks using seedlings grown on a soil infected with the fungi, some of the test compounds might be partly degraded during the test period. They considered that the steric bulk of N-substituents (normal and branched alkyls) has a role to protect the compounds against degradation.

They transposed the above consideration along with the suggestion based on Eq. 17 and 18, that the steric congestion due to aromatic ortho substituents is

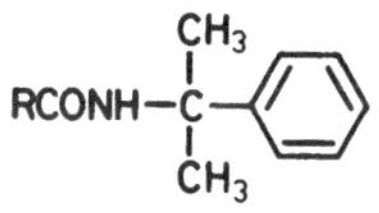

17

favorable to the herbicidal activity of N-chloroacetyl-N-phenylglycine esters, to optimization of the activity of a new class of herbicides such as N-(1-methyl-1-phenethyl)acylamide derivatives (*17*) having the amide moiety in common with the N-chloroacetylphenylglycine esters.

A systematic modification of the structure was performed to make the substituent R bulkier and bulkier to protect against the possible hydrolytic detoxication mechanism. In the course of the structural modification, the QSAR analyses were utilized to get informations for structural factors to optimize the activity. The analysis for a set of 41 derivatives where R varies from simple alkyls to such congested groups as i-Pr(Me)$_2$C and t-Bu(Cl)CH gave Eq. 34 [31],

$$pI_{50} = \underset{(0.093)}{-0.151\pi^2} + \underset{(0.457)}{0.983\pi} - \underset{(0.070)}{0.350 E_s^c} + \underset{(0.465)}{2.877} \qquad (34)$$
$$n = 41 \quad s = 0.267 \quad r = 0.933$$

where I_{50} is the molar concentration required for 50% inhibition against shoot elongation of the seedlings of bulrush, *Scirpus juncoides*, after 12 days. Eq. 34 indicates that, although there is an optimum in hydrophobicity within the set of 41 compounds ($\pi_{opt} = 3.3$), the steric bulkiness of the R substituent in terms of E_s^c is still to be augmented for the higher activity.

Further structural modifications were made on the α-bromoacylamide structure (*18*). For 14 derivatives where R′ is varied from simple alkyl to such highly branched

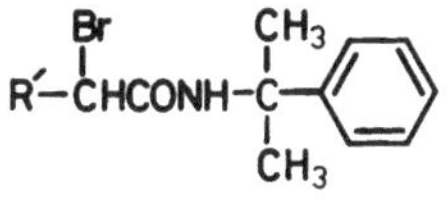

18

groups as i-Pr(Me)$_2$C, i-Bu(Me)$_2$C and (Et)$_2$MeC, Eq. 35 was derived to show that the optimum for the steric effect of R′ exists at about −3.00 in terms of E_s^c [31].

$$\begin{aligned} pI_{50} &= -\underset{(0.152)}{0.199}(E_s^c)^2 - \underset{(0.546)}{1.227}E_s^c + \underset{(0.364)}{4.393} \\ n &= 14 \quad s = 0.242 \quad r = 0.940 \end{aligned} \tag{35}$$

The E_s^c value for R′ is used in Eq. 35, since the value for such highly congested groups as R′CHBr is difficult to estimate accurately. From these informations and considering a certain, but not significant, collinearity between E_s^c and π values for compounds included in Eq. 35 as well as the easiness of preparation, S-47 (*18*; R′ = t-Bu) where $\pi(R) = 2.2$ and $E_s^c(R') = -2.5$, was selected as a promising novel paddy field herbicide and forwarded to extensive field trials.

3.2.3 Inhibition of Acetylcholinesterase by Aliphatic Ammonium Ions

It is well known that the catalytic center of acetylcholinesterase comprises two types of binding site, the anionic and the esteratic sites [32]. The interaction between positively charged quaternary nitrogen of acetylcholine molecule and the anionic site of the enzyme is suggested to play an important role in the binding of the substrate with the enzyme [32]. To understand the role of this interaction, we attempted to analyze the inhibitory activity of simpler ammonium ions without the ester moiety [23]. Using bovine erythrocyte acetylcholinesterase preparation, the inhibitor constant K_i (the dissociation constant of enzyme-inhibitor complex) was determined for 50 primary, secondary, tertiary and quaternary ammonium ions. Preliminary examinations revealed that each of the four N-substituents may exert hydrophobic as well as steric effects on the interaction with the enzyme specifically depending on the classification according to the relative bulk. We classified four substituents in the same manner as for the ion-pair formation. Unfortunately, it is impossible to definitely separate hydrophobic and steric effects from each other with the use of π_i and $E_s^c(R^i)$ values for each of the four N-substituent sets, since the internal correlation between these two constants within each substituent set is fairly high.

Since the interaction of ammonium ions with the anionic site is regarded as being a sort of ion-pair formation, we expected that the total steric effect of N-substituents on the enzyme-inhibitor interaction would be quite similar to that on the ion-pair partition equilibrium with the picrate anion. We used the steric constant terms in Eq. 28, $1.049E_s^c(R^1) + 0.682E_s^c(R^2) + 0.235E_s^c(R^3)$, together as a single steric parameter, $\Sigma\delta_i E_s^c(R^i)$, so that the internal correlations between independent parameters became negligible. As expected, Eq. 36 shows that the specific hydrophobic effect of substituents is really separable with the use of the model for the steric effect.

$$\begin{aligned} \log(1/K_i) &= \underset{(0.169)}{0.545}(\pi_1 + \pi_4') + \underset{(0.229)}{1.075}\pi_2 + \underset{(0.141)}{0.392}\pi_3 + \underset{(0.214)}{0.827}\pi_4'' \\ &\quad + \underset{(0.425)}{1.152}\Sigma\delta_i E_s^c(R^i) - \underset{(0.112)}{0.195}n_H + \underset{(0.425)}{1.419} \\ n &= 50 \quad s = 0.269 \quad r = 0.942 \end{aligned} \tag{36}$$

The coefficient of the steric term is close to 1 indicating that the stereospecific fit of ammonium ions with the anionic site of acetylcholinesterase is almost identical in nature with that required for the ion-pair formation with picrate. The specific hydrophobic effect of substituents as revealed by Eq. 36 suggests that the hydrophobic nature of the enzymic milieu surrounding the anionic site is not uniform. The coefficient associated with each of the π terms probably indicates the hydrophobic nature of the enzyme surface corresponding to each of substituents. π_4' and π_4'' represent, respectively, the hydrophobicity of the main and the side chains of R^4 substituents. The non-uniform hydrophobic effect of the R^4 substituents was also ascertained by examining the K_i values for 8 primary and 8 alkyltrimethyl quaternary ions having variously branched R^4 substituents. The n_H term having negative sign probably indicates that the larger the number of NH hydrogen, the more is the association with the enzyme prevented by the hydration in the aqueous bulk phase. The hydrogen acceptor in the enzymic milieu, if any, seems to be less basic than water. However, there is a certain degree of collinearity between n_H and $\Sigma\sigma^*$ value over four N-substituents. We have to wait for the definite answer in this respect until the inhibition data for such ammonium ions as those having CF_3 substituent are determined so that the collinearity is no longer significant.

We can draw schematically the interaction of ammonium ions with the anionic site of acetylcholinesterase in Fig. 5 based upon Eq. 36 since the informations for the steric and hydrophobic effects of substituents do not change practically even if $\Sigma\sigma^*$ is used for the analysis [26].

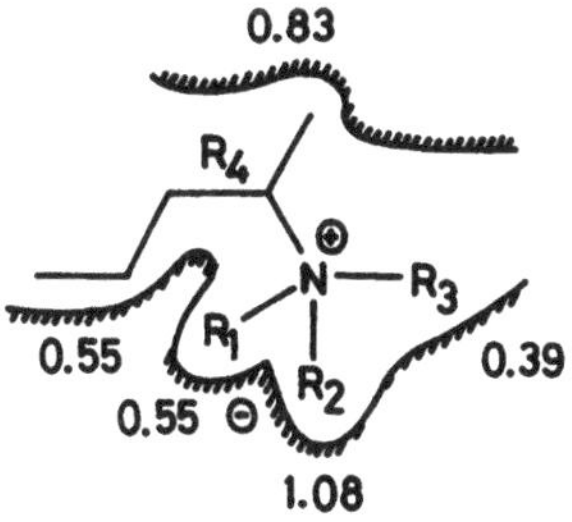

Fig. 5. Susceptibility of the binding vicinal to the anionic site of acetylcholinesterase to the hydrophobicity of N-substituents [26] (reproduced with permission from Pure and Applied Chemistry, IUPAC)

4 STERIMOL Parameters

4.1 Free-Energy-Related Background of STERIMOL Parameters

STERIMOL parameters developed by Verloop et al. are a set of parameters for a substituent representing directional nature of steric dimensions [33]. For each substituent, an axis connecting the α-atom with the rest of the molecule is first defined. The value of the length parameter, L, is simply the length of the substituent in Å along the direction of this axis. The substituent is then projected on a plane perpendicular to the L-axis. Among tangential lines drawn on the projection, the one, the distance of which is shortest from the axis, is then selected. The shortest distance is defined as the shortest width parameter, B_1. The three other width parameters, B_2, B_3 and B_4, in ascending order of the value, are then defined as the distance to the paired parallel tangential lines on the projection as shown in Fig. 6. For conformationally flexible substituents containing hydrocarbon chains, parameters are usually estimated on the basis of fully extended structure.

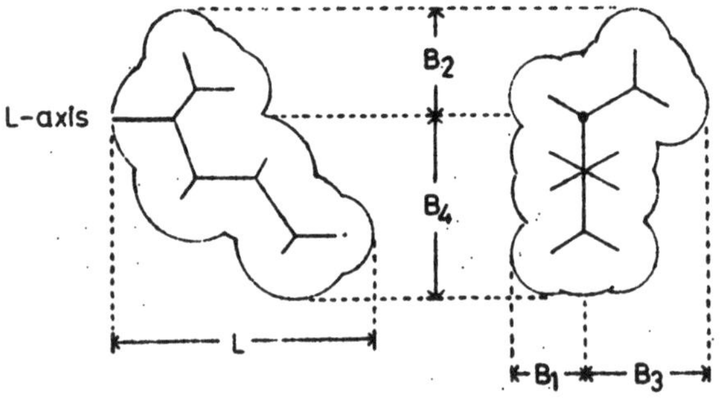

Fig. 6. Definition of STERIMOL parameters as for the 2-pentyl group having fully extended staggered conformation

Although these parameters represent the shape of substituents as a set, there is some collinearity among B_1, B_2 and B_3 for a number of substituents. Thus, B_1, B_4 and L, as the most independent variables, are usually used in QSAR analysis [33]. More recently, Verloop proposed the maximum width parameter, B_5, and showed that, in most cases, B_5 works well in place of B_4 [35]. Since they are defined mechanically as the length or width, the background of their utilization along with other free-energy related parameters is not necessarily clear.

We have recently studied this point by examining the steric effect of meta-substituents of phenyl acetate on the reactivity with α- and β-cyclodextrins (CD) as enzyme model [35]. Cyclodextrins form an inclusion complex with phenyl acetates (PA) and then catalyze the cleavage of the ester linkage [36] as shown in Eq. 37.

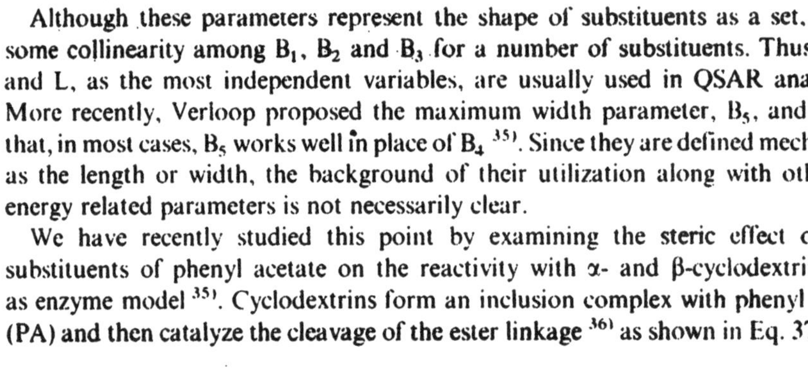

(37)

Eq. 38 and 39 were derived for the inclusion complex formation, $1/K_d$ (= k_1/k_{-1}), with β- and α-cyclodextrin, respectively.

$$\log(1/K_d) = \underset{(0.101)}{0.957\pi} + \underset{(0.107)}{1.841} \tag{38}$$
$$n = 16 \quad s = 0.129 \quad r = 0.983$$

$$\log(1/K_d) = \underset{(0.218)}{0.591\pi} + \underset{(0.573)}{1.310B_1} - \underset{(0.880)}{0.445} \tag{39}$$
$$n = 14 \quad s = 0.251 \quad r = 0.943$$

The π values used here are derived from di-n-butylether/water partition coefficients since the ether phase was thought to better simulate the environment of the cavity of cyclodextrins. Eq. 38 indicates that the principal driving force for the inclusion complex formation with β-cyclodextrin is the dehydration from surroundings of the guest molecule.

In Eq. 39, substituents, the dimension of which is larger than 5.1 Å in terms of the sum of the opposite-pair B_i values, are not included, since their $\log(1/K_d)$ values are much lower than those expected from the correlation for others. They are supposed not to be accomodated into the cavity of the α-cyclodextrin, the diameter of which is 4.5-5 Å. Thus, the inclusion complex formation with α-cyclodextrin is more limited than that with β-cyclodextrin and the extent of dehydration is less significant. The term B_1 in Eq. 39 implies that the larger the minimum width of substituents the maximum diameter of which is less than 5.1 Å, the more uniform would be the van der Waals interaction with the cavity wall resulting in the more stable inclusion complex.

Eq. 40 and 41 were formulated for the catalytic process, $\log k_2$, with β- and α-cyclodextrin, respectively.

$$\log k_2 = \underset{(0.247)}{1.283\sigma^0} + \underset{(0.058)}{0.275 D_{max}} - \underset{(0.278)}{3.496} \tag{40}$$
$$n = 19 \quad s = 0.123 \quad r = 0.959$$

$$\log k_2 = \underset{(0.323)}{1.445\sigma^0} + \underset{(0.441)}{0.878 D_{max}} - \underset{(0.051)}{0.103 D_{max}^2} - \underset{(0.915)}{3.671} \tag{41}$$
$$n = 21 \quad s = 0.158 \quad r = 0.951$$

In these equations, D_{max} is the larger one of the summed values of B_i parameters of the opposite pair. Eq. 40 indicates that the larger the "maximum diameter" within the range of substituents studied, the more properly they fit the edge of the cavity

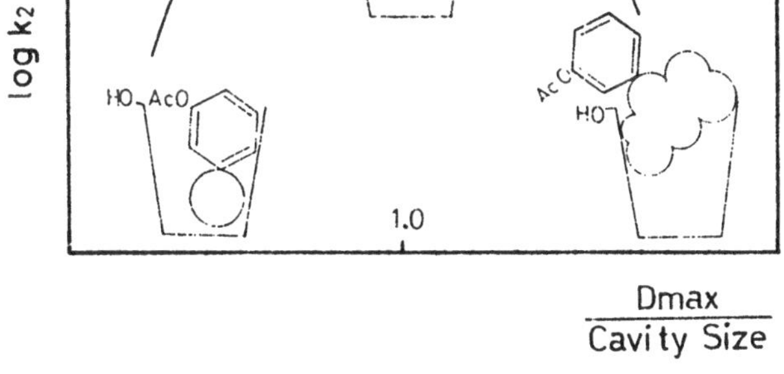

Fig. 7. Parabolic relationship between reactivity with cyclodextrin and substituent "diameter" of m-substituted phenyl acetates

of β-cyclodextrin, the diameter of which is ca. 7 Å. Thus, the easier will be the nucleophilic attack of the peripherally situated 2- or 3-O^- ion against the carbonyl carbon of the substrate side chain to initiate the hydrolysis.

Eq. 41 suggests the existence of an optimum "diameter" for the proper fit of m-substituents in the cavity of α-cyclodextrin. The optimum D_{max} value is estimated as 4.4 Å from Eq. 41 which is approximately equivalent to the diameter of the cavity of α-cyclodextrin. The situation is shown in Fig. 7.

After separating steric effect by the STERIMOL term, the slope of the σ^o terms in Eq. 40 and 41 is close to those observed for the same type of acylation reactions in a number of physical organic systems [37]. Thus, STERIMOL values can well be used as "free-energy-related" parameters in combination with σ and π parameters and provide informations about three-dimensional aspects of the interactions between drugs and biomolecules.

4.2 Applications of STERIMOL Parameters

4.2.1 Applications to Development of Pesticides

Because benzanilides exhibit fungicidal activity [38] and anthranilic acid esters have fungicidal and bactericidal activities [39,40], Kirino et al. attempted to derive a new class of fungicides from those having a hybrid structure, methyl N-(substituted benzoyl)anthranilates [41]. Among various derivatives, m-substituted compounds (*19*) were shown to exhibit appreciable preventive activity against powdery mildew of cucumber.

$$\begin{aligned} pI_{50} &= -0.114L^2 + 1.082L - 0.715B_4 - 3.349 \\ &\qquad (0.104) \qquad (0.839) \qquad (0.291) \qquad (1.640) \qquad\qquad (42) \\ n &= 14 \quad s = 0.229 \quad r = 0.942 \end{aligned}$$

In Eq. 42, I_{50} is the molar concentration required to inhibit the disease development 50% of the control.

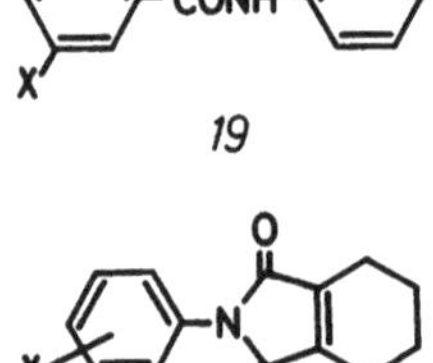

19

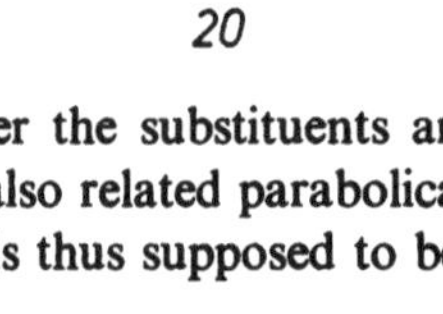

20

The wider the substituents are in the direction B_4, the lower the activity. The activity is also related parabolically to the substituent length. The steric fit with the target site is thus supposed to be critical for activity. The compounds such as m-F,

-Cl and -CN with smaller B_4 values and lengths as near as possible to the optimum, exhibited high potency as expected.

In the course of developing herbicides possessing N-phenyl tetrahydrophthalimide structure (*20*), Ohta et al. investigated structural requirements for the activity against sawa millet, *Echinochloa utilis*, of m- and p-substituted derivatives[42]. As shown by Eq. 43, the steric dimensions expressed by the STERIMOL length parameter L_p and "maximum" width B_4^p of p-substituents play a decisive role in determining activity.

$$pI_{50} = \underset{(0.348)}{-0.597\sigma} + \underset{(0.316)}{1.768L_p} - \underset{(0.084)}{0.312L_p^2} - \underset{(0.212)}{0.946B_4^p} + \underset{(0.217)}{4.067} \quad (43)$$
$$n = 28 \quad s = 0.273 \quad r = 0.930$$

I_{50} is the 50% inhibitory molar concentration against the root growth.

The lack of significance of the hydrophobic parameter term seems to mean that translocation to the target site is not critical among the compounds examined. Together with this, the significance of steric effects only for p-substituents suggests that the substituent effects are primarily related to the interaction with the critical site of herbicidal action. In this respect, hydrogen bond formation at the carbonyl oxygen atom with an acidic group at the site of action is a possibility, since the negative sign of the σ term shows that the electron-withdrawing effect of the substituents directed to the imide moiety is unfavorable to activity.

4.2.2 Cytokinin-active Adenine Derivatives

Cytokinins are a class of plant hormones which control cell division and growth. Following the discovery of kinetin (N^6-furfuryladenine), a large number of N^6-

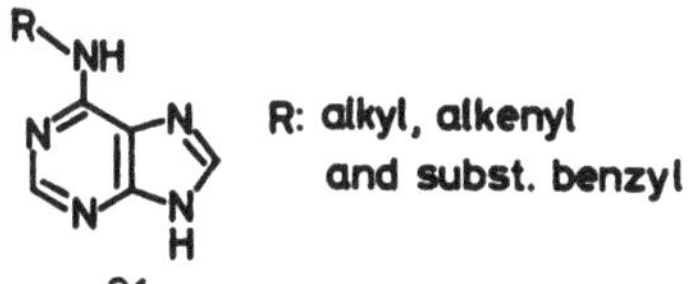

21

substituted adenines and related compounds have been synthesized[43]. We have prepared a number of N^6-alk(en)yl- and N^6-substituted benzyl derivatives (*21*) and performed QSAR analysis separately for these two sets to give Eq. 44 and 45[44], N^6-Alkyladenines:

$$\log(1/E_{50}) = \underset{(0.29)}{-0.56B_4^2} + \underset{(3.04)}{5.84B_4} + \underset{(1.02)}{1.66\sigma^*} - \underset{(7.81)}{7.67} \quad (44)$$
$$n = 12 \quad s = 0.20 \quad r = 0.91$$

N^6-Benzyladenines:

$$\log(1/E_{50}) = \underset{(0.50)}{-0.94B_4} - \underset{(0.35)}{0.71B_{o,m}} + \underset{(3.32)}{13.41} \quad (45)$$
$$n = 11 \quad s = 0.27 \quad r = 0.87$$

where E_{50} is the molar concentration at which the 50% callus yield of the maximum response is given in the tobacco callus bioassay. B_4 is estimated for the fully extended conformation of N^6-substituents. The $B_{o,m}$ is the "maximum" width for o- and m-substituents of N^6-benzyl moiety, the bulkiness of which is not accounted for by the B_4 for whole of the N^6-benzyl substituents.

The significance of the term B_4^2 in Eq. 44 indicates that there is an optimum steric condition (ca. 5.2 Å) for N^6-alkyl groups in terms of the "maximum" width to exhibit the activity. The B_4 value for the benzyl substituents is always higher than the optimum B_4 value for the alkyl groups, even that for the unsubstituted benzyl being 6.02. Thus, the negative sign of the term B_4 in Eq. 45 suggests that the activity of benzyl derivatives is on the supraoptimal downward part of a parabolic relationship with the B_4 value which is common for the two series. The analysis of the combined set of compounds gave Eq. 46.

$$\log(1/E_{50}) = -\underset{(0.15)}{0.32}\,B_4^2 + \underset{(1.71)}{3.35}\,B_4 - \underset{(0.26)}{0.65}\,B_{o,m} + \underset{(0.98)}{2.03}\,\sigma^* - \underset{(4.74)}{1.50} \qquad (46)$$

$$n = 22 \qquad s = 0.26 \qquad r = 0.85$$

The σ* value is that for N^6-substituents indicating an electron withdrawing effect directing to the N^6-imino group. The insignificant term σ* in Eq. 45 is due to a narrower range of its value in benzyl derivatives. The electronic interaction at the N^6–H with a basic site on the receptor is possible. No hydrophobicity seems to play a significant role in the transport and/or binding processes. Variations in the cytokinin activity are governed mainly by the variations in the interaction with the target site of action, where the accomodation to the receptor cavity appears to be the predominant factor and the electronic effect intensifies the binding further.

4.2.3 Cytokinin-active Diphenylureas

N,N'-Diphenylurea is a compound isolated and identified as a cell-division factor of coconut milk which exerts the same activity as the N^6-substituted adenines in the cytokinin tests [45]. Bruce and Zwar synthesized a number of ring substituted derivatives and determined their activity [46]. We tried to analyze quantitatively their data for derivatives where one of the benzene rings is unsubstituted (*22*) [44].

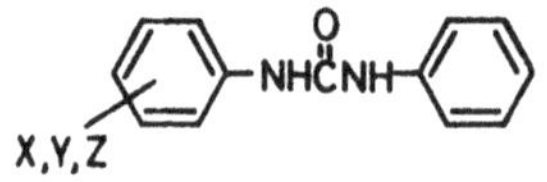

22

Preliminary analyses indicated that di- and tri-substituted compounds can be incorporated into the relationships found for each of o-, m- and p-monosubstituted derivatives according to the following principles:

a) Polysubstituted compounds having a substituent at the o-position are assigned to be the "ortho" derivatives regardless of substitutions at the other positions.
b) Those having substituents at both the m- and p-positions are classified as "para" compounds.

c) The rest, those having substituents only at one or two of the m-positions, are classified as the "meta" series.
d) The position-specific steric and hydrophobic effects are considered only for the single substituents located at each assigned position.

Position-specific substituent effects were analyzed by Eq. 47,

$$\log(1/C) = \underset{(0.33)}{0.90\sigma} - \underset{(0.25)}{0.85\,L_o} - \underset{(0.22)}{0.27\,L_p} + \underset{(0.58)}{1.04\pi_m} + \underset{(0.32)}{5.00} \qquad (47)$$
$$n = 39 \quad s = 0.38 \quad r = 0.91$$

where C is the minimum molar concentration that gives a detectable response in the tobacco pith assay. L_o and L_p are the STERIMOL length parameter for o- and p-substituents, respectively. The negative slope of the term L_o greater than that of the term L_p indicates the more strict steric demand at the ortho position. The high hydrophobicity of the m-substituents enhances activity whereas the steric effect is insignificant at this position. These results rationalize and provide physicochemical basis for the recognized general order of potency among isomers: *meta* > *para* > *ortho*.

4.2.4 Pyridopyrimidine Anticytokinins

We have developed recently a couple of classes of anticytokinins which antagonizes the effect of cytokinins [47,48]. The anticytokinin activity of N^4-substituted 4-amino-2-methylthiopyrido[2,3-d]pyrimidine derivatives (*23*), has been analyzed to give Eq. 48 [48].

$$pI_{50} = -\underset{(0.29)}{0.96B_4^2} + \underset{(2.67)}{8.75B_4} - \underset{(0.43)}{0.58\pi} - \underset{(5.86)}{12.32} \qquad (48)$$
$$n = 10 \quad s = 0.28 \quad r = 0.96$$

where I_{50} is the molar concentration required for the 50% inhibition against the tobacco callus growth under control conditions with kinetin but without anticytokinin.

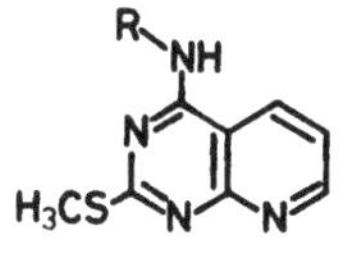

23

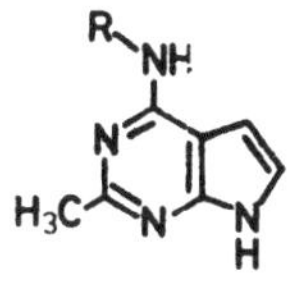

24

The optimum B_4 value for N^4-substituents, 4.5 Å coincides very well not only with that for N^6-substituents of adenylate cytokinins, 5.2 Å, shown in the previous section, but also with the value of 4.7 Å observed for another class of compounds, N^4-substituted 4-amino-2-methylpyrrolo[2,3-d]pyrimidines (*24*) [47] which show either cytokinin or anticytokinin activity depending upon the structure of N^4-substituents. These results in combination may help to think of the size of structure which should sterically fit well the cytokinin receptor cavity.

The variation of the activity from agonistic to antagonistic in the pyrrolo[2,3-d]-pyrimidine derivatives is consecutive in terms of structure and well elucidated by the B_4 value of the N^4-substituents [47]. The B_4 value within a range of 4.7–6.0 yields agonists, whereas that outside this range leads to anticytokinins. Steric dimensions of compounds are important factor determining the type of the activity as well as the potency.

4.2.5 Perillartine and Related Synthetic Sweetners

Perillartine (*25*) is the *syn*-oxime of perillaldehyde isolated from *Perilla arguta* Benth., and was found to be about 2000 times as sweet as sucrose in 1920 in Japan [49]. Acton and Stone prepared a number of related aldoximes (*26*) and determined their taste potency [50,51]. In quantitative analysis of their data, we have utilized steric parameters

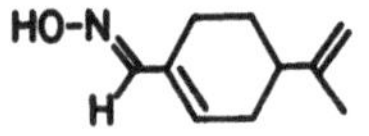

25

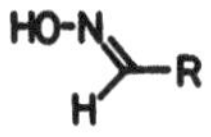

26

for substituent R (in *26*) slightly modified from STERIMOL values [52]. They are shown schematically in Fig. 8. W_l is taken as the width in the direction to which

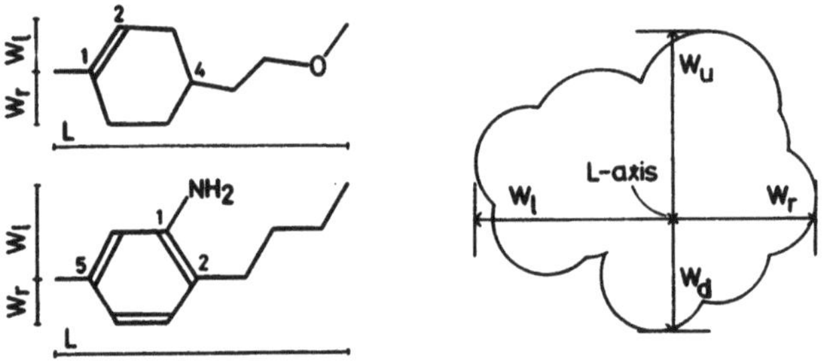

Fig. 8. Definition of modified STERIMOL parameters; the upper is for perillartine and related oximes and the lower is for substituted anilines [52] (reproduced with permission from the American Chemical Society)

the 4-substituent extends in the fully extended conformation. The $>C^1=C^2<$ double bond is assumed to exist in the same direction. W_r is the width in the direction opposite to W_l. W_u and W_d are the thickness upward and downward, respectively.

Eq. 49 is the best correlation derived [52],

$$\log A = \underset{(0.31)}{0.63} \log P + \underset{(0.10)}{0.19} L - \underset{(0.18)}{0.48} W_l - \underset{(0.24)}{0.26} W_u + \underset{(0.78)}{2.87} \qquad (49)$$
$$n = 38 \quad s = 0.32 \quad r = 0.91$$

where A is the taste potency, irrespective of whether the taste is sweet or bitter, relative to that of sucrose. The positive log P term perhaps reflects the partitioning process from saliva onto the receptor site of the tongue. The negative coefficients of the W_l and W_u terms indicate that the molecular width and thickness are detrimental to the taste potency. The positive L term means that longer molecules are favorable to the activity within the set of 38 compounds.

Acton and Stone also determined the sweet/bitter ratio of the compounds[51]. Taking the compounds with the sweet/bitter ratio >1.00 as sweet, the maximal contour viewed from the α-atom of the R moiety to the L-direction of the all

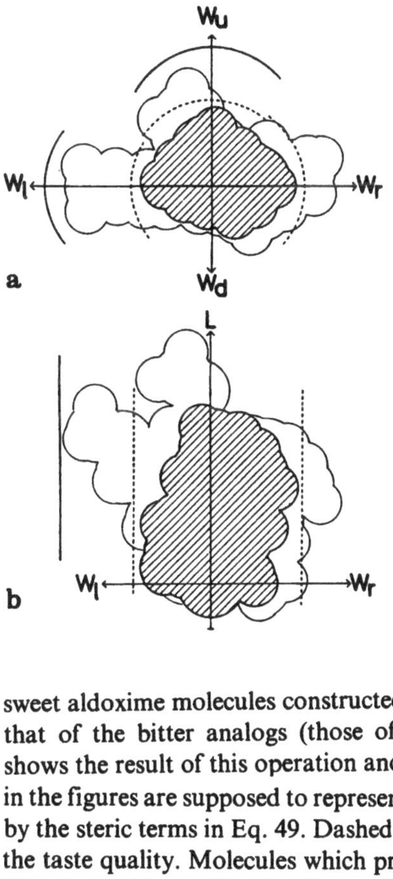

Fig. 9a and b. Aldoxime-receptor binding model. The maximum contour of the sweet aldoximes is striped. Unfigured is that of the bitter analogs. **a:** The view from the C^1 atom to the direction of L-axis. **b:** The view from the upside.[52] (reproduced with permission from the American Chemical Society)

sweet aldoxime molecules constructed by the CPK model was drawn and overlapped that of the bitter analogs (those of which the sweet/bitter ratio <1.00). Fig. 9a shows the result of this operation and Fig. 9b is the view from the upside. Solid lines in the figures are supposed to represent the spatial walls for the taste potency indicated by the steric terms in Eq. 49. Dashed lines are suggested to be barriers discriminating the taste quality. Molecules which project from the striped, sweet contour are bitter. They have either a W_r value larger than 5.8, a W_u value larger than 3.5 or a W_l value larger than 3.5. The W_r value is not significant in determining the taste potency so far as Eq. 49 is concerned but plays an important role in controling the taste quality.

It is interesting to note that similar steric conditions are required for sweetness of nitro- and cyano-anilines. Using the W parameters defined similarly to those of aldoximes as shown in Fig. 8, the combined data for fourteen 2-substituted 5-nitroanilines and six 2-substituted 5-cyanoanilines were analyzed to give Eq. 50[52].

$$\log A = 0.52L - 1.37W_1 + 3.71 \qquad (50)$$
$$(0.14) \quad (1.08) \quad (3.49)$$
$$n = 20 \quad s = 0.32 \quad r = 0.90$$

Since no bitter compounds are reported, the A value in Eq. 50 is the net sweet intensity relative to sucrose. The nitro and cyano groups seem to play an identical role and the substituent effects are common in these two series of compounds.

4.2.6 L-Aspartyl Dipeptide Sweetners

The first member of this class of sweetners, L-aspartyl-L-phenylalanine methyl ester (*27*), was found by chance by Mazur et al., when synthesizing gastrin C-terminal tetrapeptide, tryptophylmethionylaspartylphenylalaninamide [53]. Since then, extensive studies have delineated structure-activity relationship of related dipeptides and

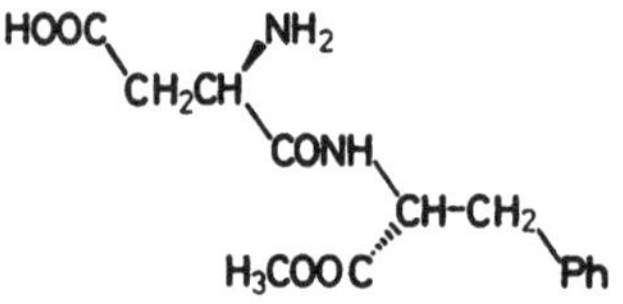

27

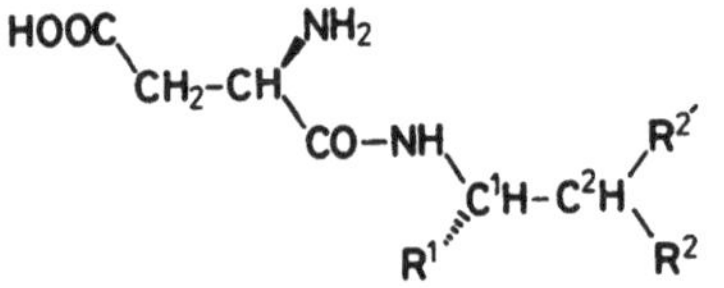

28

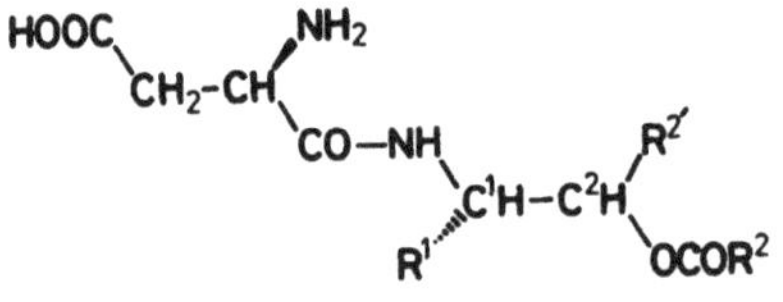

29

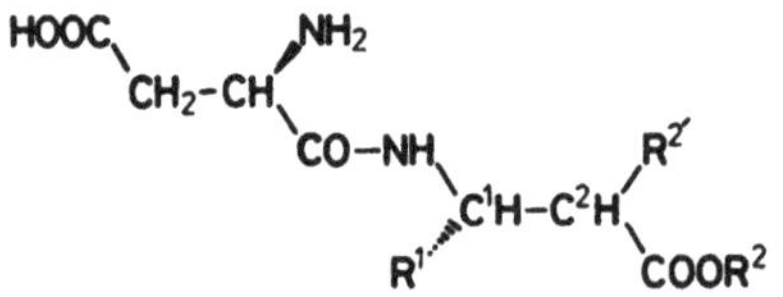

30

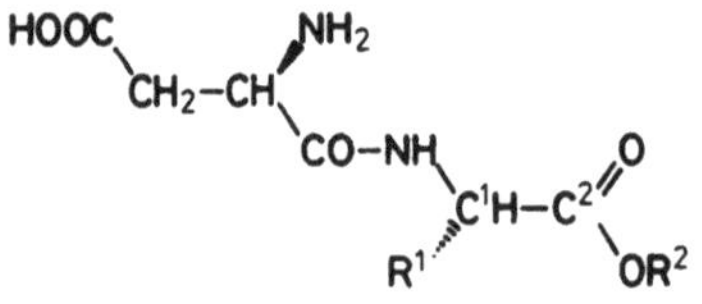

31

analogs [54-60]. They are classified into the following four types: L-aspartic acid amides (*28*), L-aspartylaminoethyl esters (*29*), L-aspartylaminopropionates (*30*) and L-aspartylaminoacetates (*31*). Our quantitative analyses were performed systematically according to the above structural classification and the importance of the steric dimensions in terms of the modified STERIMOL parameters were disclosed [61].

Eq. 51 was formulated for the data of aspartylaminoacetates [56–59],

$$\begin{aligned}\log A = {} & \underset{(0.51)}{1.54\sigma^*} + \underset{(0.61)}{1.49L_1} - \underset{(0.07)}{0.19L_1^2} + \underset{(1.04)}{2.46L_2} - \underset{(0.09)}{0.21L_2^2} \\ & + \underset{(0.25)}{1.05(W_1)_2} + \underset{(0.29)}{0.37I_1} - \underset{(0.29)}{0.74I_2} - 10.87 \qquad (51) \\ n = 56 \quad & s = 0.34 \quad r = 0.95\end{aligned}$$

where A is the sweet potency relative to sucrose. The A value of this series of compounds varies in a range of 10^4-fold, the widest range among others. Substituent R^1 in *31* is taken so that its length parameter L_1 is always lower than that of the other substituent, $COOR^2$. L_2 is the length parameter for OR^2. The parabolic relationship with L_1 as well as L_2 shows that optimum length of the R^1 and OR^2 substituents is about 6 and 4, respectively. The sweetness of the cyclohexanol esters (*31*: R_2 = subst. cyc-hexyl), having at least one methyl group at the 2-position of the cyclohexane ring, is reportedly very high [59], being $1\text{–}6 \times 10^4$ times sweeter than sucrose. This structural characteristic is manifested by a width parameter, $(W_1)_2$, when it is measured from C^1—C^2 bond axis of the alcohol moiety as depicted in Fig. 10.

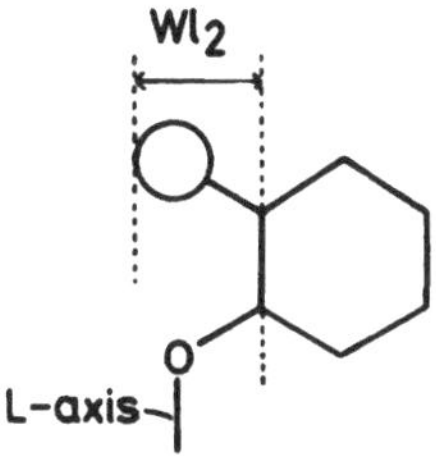

Fig. 10. $(W_1)_2$ parameter used in Eq. 51. The figure depicts the OR^2 moiety of the cyclohexyl esters of L-asparthyl-aminoacetic acid (*31*) [61] (reproduced with permission from the American Chemical Society)

σ^* is taken for the N-substituents, $—C^1H(R^1)COOR^2$, as the sum of $\sigma^*(CH_2R^1)$ and $\sigma^*(CH_2COOR^2)$, which were estimated according to $\sigma^*[(CH_2)_nX] = \sigma^*(X) \times 0.36^n$ where 0.36 is the transmission factor and X is heterosubstituents [4]. The positive coefficient of this term indicates that the electron withdrawing effect of N-substituents on the Asp-NH enhances sweet potency.

I_1 is an indicator variable which takes a value of 1 for the data determined by Mazur et al. to correct parallel difference from those by Ariyoshi et al. Those by Fujino et al. need no correction. Examination of the data by Ariyoshi et al. indicated that there is a parallel difference in the sweetness between compounds derived from D-threonine and the allo counterparts [*31*: R^1 = CH(OH)Me]. I_2 is the indicator variable taking care of the allo compounds. The negative coefficient of this term indicates that the allo configuration within the R^1 moiety is sterically unfavorable.

For other series of aspartyl sweetners, similar quantitative correlations were derived [61]. In each series, steric dimensions of the C^1 and C^2 substituents in terms of length and/or width parameters are crucial determinants for the sweet potency.

The positive σ* term is common among four series of sweetners. The coefficient, 1.54, in Eq. 51 is, however, about twice those for other series which is about 0.7. This suggests that the aspartylaminoacetates (*31*) binds tighter with the basic site of the receptor perhaps by the hydrogen bonding than do the other three classes of compounds.

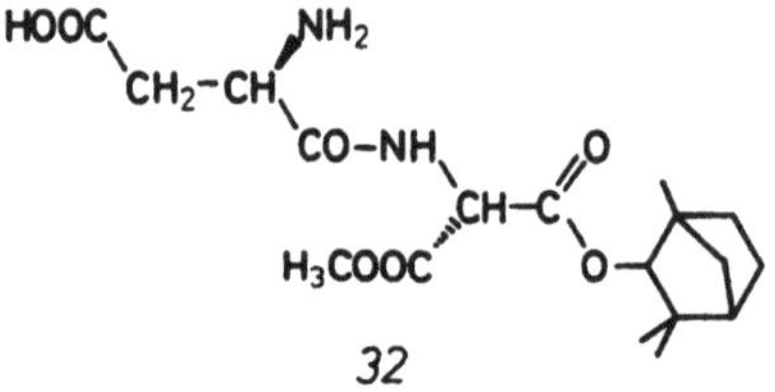

32

The sweetest derivative included in Eq. 51 is L-aspartylaminomalonic acid methyl fencyl ester (*32*) which is 63000 times as sweet as sucrose [59]. This is understandable on the basis of the high coefficient of the σ* term of Eq. 51, a high σ* value itself of the N-substituent containing two ester carbonyl groups (estimated as 1.38) and the $(W_1)_2$ term due to the presence of methyl groups on the R^2 cyclohexyl moiety.

5 Van der Waals Molecular Volume

5.1 Scope of the van der Waals Volume

Van der Waals molecular volume is the volume contained by van der Waals surface of a molecule which is defined as the surface of the intersection of spheres each of which is centered at the equilibrium position of the atomic nucleus with van der Waals radius of each atom [62]. Since the van der Waals radius of an atom is the distance at which the repulsive force balances the attraction forces between two non-bonded atoms, van der Waals molecular volume is regarded as the volume impenetrable for other molecules with thermal energies at ordinary temperatures.

Van der Waals volume of a molecule or a substituent at a certain position in the molecule has been utilized in QSAR studies quite successfully for a number of cases [63,64]. By intuition, this parameter is thought to represent a type of steric effect which is volume-dependent. There have been quite a few controversies as to whether the van der Waals volume is really the steric parameter. Charton suggested that the "volume" parameter is not likely to be a measure of the steric effect as the term used in physical organic chemistry [65], since steric effects observed in simple organic reactions are directionally dependent for nonsymmetric substituents. He further proposed that even steric effects involved in the interaction with a pocket-like receptor site in biological systems should be rationalized in terms of the widths and lengths of various directions of molecules or substituents. Since there is a certain degree of collinearity with molecular refractivity, Charton rather considered that the "volume parameter" represents polarizability which accounts for van der Waals attraction forces between biologically active compound and the receptor [65].

Leo et al. indicated that the van der Waals volume is linearly related to hydrophobicity for non-polar compounds expressed in terms of log P (octanol/water) [66]. Moriguchi et al. showed that the log P value is generally factored into two components attributable to hydrophilic effect of polar group and hydrophobic effect due to the net molar volume [67]. Thus, the van der Waals volume could be a parameter related to solute-solvent interactions and partition coefficient.

In spite of these arguments, the energy required for the creation of a cavity, the volume of which is ΔV, on biomacromolecule so as to include a drug molecule could be expressed as $p\Delta V$, p being the pressure which could be taken as a constant under physiological conditions. The volume of the cavity should correspond closely with the impenetrable volume of the engulfed part of the molecule. As Pearlman suggested [62], the biological activity of certain compounds could well be related to the volume of the substituent which interacts with the receptor site of a restricted volume.

Quite recently, Hopfinger has been developing a procedure [68] where three dimensional molecular shape is parameterized on the basis of the common steric overlap volume sharing with a reference compound having an "active" conformation analyzed using molecular mechanics. The molecular shape descriptors have been successfully applied along with other free-energy-related physicochemical parameters to QSAR analyses for dihydrofolate reductase inhibitors [69,70] and phenyltriazene anticancer compounds [71]. Although his descriptors are not necessarily of equivalent significance with the van der Waals volume parameter, his analyses seem to warrant the use of V_w for the volume-dependent steric effect. In the following examples, the van der Waals volume parameter (scaled by 0.1 to make it more nearly equiscalar with π and σ parameters) is according to Bondi [72], although it is not accurate enough neglecting overlaps of the van der Waals spheres of the types shown in Fig. 11.

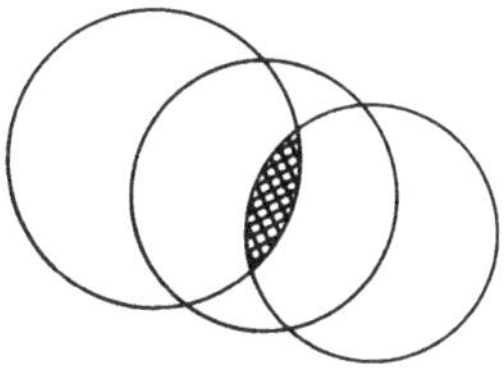

Fig. 11. Overlap (cross-striped part) of van der Waals spheres neglected in estimation of the van der Waals volume according to Bondi [72]

5.2 Applications of van der Waals Volume

5.2.1 Insecticidal Activity of Lindane Analogs

Lindane, the γ-isomer of HCH, is a potent insecticide. Its insecticidal action is due to a hyperexcitatory effect on the insect's central nervous system induced by accumulation of acetylcholine in the synaptic region [73]. We prepared a number of analogs in which some of the chlorine atoms are replaced by other substituents while maintaining the γ-configuration and determined their insecticidal activity against mosquitos [74].

Since the activity seemed highly dependent on the positions of "hetero-atom" substituents, the compounds were stereochemically classified into two groups, *meso*- and *dl*-analogs, according to their hetero-substituent positions as shown in Fig. 12. QSAR analyses for insecticidal activity of these groups were performed separately to give Eq. 52 and 53 [74].

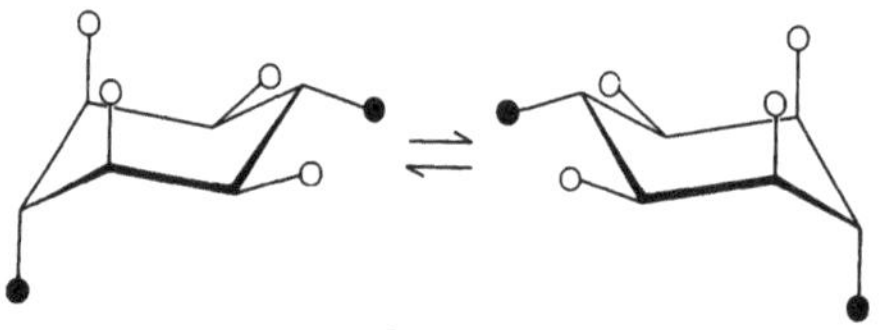

Fig. 12. Classification of substituents in cyclohexane derivatives

meso-Analogs:

$$\log(1/LD_{50}) = \underset{(0.67)}{1.66\Delta r_w} + \underset{(0.19)}{3.53} \quad n = 9 \quad s = 0.205 \quad r = 0.911 \tag{52}$$

dl-Analogs:

$$\log(1/LD_{50}) = \underset{(0.60)}{-2.10\Delta V_w^2} - \underset{(0.34)}{1.01\Delta V_w} + \underset{(0.31)}{3.09} \quad n = 16 \quad s = 0.350 \quad r = 0.920 \tag{53}$$

The Δr_w and ΔV_w are the van der Waals radius of the atom of substituents directly bound to the ring and the van der Waals molar volume of substituents, respectively, those of the chlorine substituent being taken as reference.

The variation in activity is governed mainly by the steric effect of substituents within the compounds examined in both series. However, the mode of interactions seems to be different, as revealed by the different steric parameters incorporated. The significance of the Δr_w in Eq. 52 may be that the *meso*-substituents interact in such a manner that they come into contact with the receptor surface of walls, whereas that of the ΔV_w in Eq. 53 may indicate the importance or fitting into a cavity or pocket on the receptor surface in the region where the *dl*-substituents direct. The ΔV_w value for Cl(0.0) is nearest to the optimum steric condition ($\Delta V_w \simeq -0.24$) calculated from Eq. 53. Thus, *dl*-Cl substituents in lindane are thought to be those which best fit the receptor cavity among the *dl*-substituents.

5.2.2 Neurophysiological Activity of Benzyl Chrysanthemates

The pyrethroids, a class of insecticides, are outstandingly safe to mammals and have been widely used against the household pest insects. A number of classical as well as recently developed synthetic analogs having high potency are esters of chrysanthemic acid with substituted benzyl alcohols. The principal target site of this class of compounds is believed to be the axonal membrane of the nervous system of insects [75]. Recently, we have determined the activity to induce a hyperexcitatory symptom against the excised nerve cords of the American cockroaches immersed in physiological saline solution containing various concentrations of substituted benzyl (+)-trans-chrysanthemates (*33*) in terms of the minimum concentration (MEC in M)

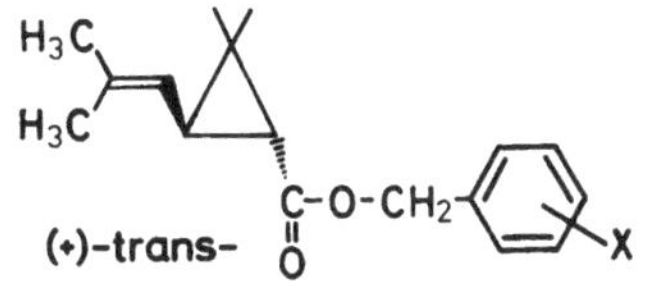

33

required to induce the repetitive train of impulses in response to a single stimulus [76]. The quantitative analysis has been performed for the ortho, meta and para substituted derivatives separately to yield Eq. 54, 55 and 56 [77].

For *ortho* derivatives,

$$\log(1/\mathrm{MEC}) = \underset{(0.480)}{5.501} - \underset{(0.673)}{1.011\sigma} - \underset{(0.332)}{0.301\pi^2} + \underset{(0.332)}{0.405\Delta V_w} \qquad (54)$$
$$n = 14 \quad s = 0.372 \quad r = 0.878$$

For *meta* derivatives,

$$\log(1/\mathrm{MEC}) = \underset{(0.459)}{5.454} + \underset{(0.397)}{0.654\Delta V_w} - \underset{(0.066)}{0.066\Delta V_w^2} \qquad (55)$$
$$n = 17 \quad s = 0.325 \quad r = 0.850$$

For *para* derivatives,

$$\log(1/\mathrm{MEC}) = \underset{(0.578)}{5.775} - \underset{(0.205)}{0.344\pi} + \underset{(0.522)}{1.132\Delta V_w} - \underset{(0.098)}{0.260\Delta V_w^2} \qquad (56)$$
$$n = 18 \quad s = 0.398 \quad r = 0.933$$

ΔV_w in this and the following section means the value relative to that of H.

Eq. 54 indicates that the larger the van der Waals volume, the more favorable are the *ortho* substituents to the excitatory activity on the nerve cords within substituents tested here. The electron-withdrawing effect of the ortho substituents deteriorates the activity. The activity varies parabolically with the π value, the optimum of which is around $\pi = 0$. Eq. 55 and 56 show that the optimum van der Waals volume exists at about 4.9 and 2.2 for meta and para substituents, respectively. The hydrophobicity of substituents is not favorable to the activity at the para position.

In a concentration range higher than those exhibiting repetitive responses, this class of compounds blocks the nerve conduction. We have also determined neuroblocking activity in terms of the minimum effective concentration (MBC in M) in the saline solution [76]. In contrast to effect on excitatory activity, substituent effects on blocking activity were not specific to substituent positions. With the position-independent hydrophobic and steric parameters, Eq. 57 is formulated for 20 compounds where finite activity indices are available [77].

$$\log(1/\text{MBC}) = \underset{(0.240)}{3.972} + \underset{(0.138)}{0.370\Delta V_w} - \underset{(0.158)}{0.283\pi^2} \qquad (57)$$
$$n = 20 \quad s = 0.284 \quad r = 0.812$$

Peculiar "topographical" effects of substituents have long been observed for the toxicity of substituted benzyl chrysanthemates against house flies [78]. The effect of the benzyl group attached to the benzyl ring is highest in the meta while lowest in the ortho derivative. For the allyl derivatives, activity was highest in the para and lowest in the ortho isomer. Eq. 54–56 indicate that the optimum van der Waals volume of substituents for the neuro-excitatory activity is largest at the ortho and smallest at the para position. In Fig. 13, the neuroexcitatory activity is expressed

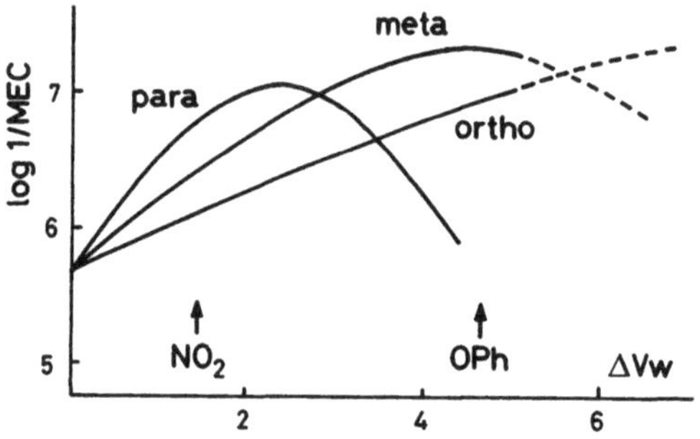

Fig. 13. Position-dependent steric effect on the neuroexcitatory activity of substituted benzyl (1R)-trans-chrysanthemates [77] (reproduced with permission from Academic Press, Inc.)

as the function of the ΔV_w value at each position according to Eq. 54–56. It is easily understood that, other factors being equal, the sequence of the activity among positional isomers varies from *para* > *meta* > *ortho* via *meta* > *para* > *ortho* to *meta* > *ortho* > *para* with increase in the bulkiness of substituent from nitro to phenoxy. Since the neuroblocking activity is not position-specific, the peculiar topographic effects of substituents on the insecticidal activity are understood on the basis of those on the neuroexcitatory activity.

5.2.3 Neuroexcitatory Activity of DDT Analogs and Related Compounds

Insecticidal activity of DDT and related compounds is due to their neuroexcitatory effect, the mechanism of which is very similar to that of pyrethroids [75]. Using a number of analogs of DDT (*34*), DDD (*35*), and prolan (*36*), p,p′-substituents of which are modified variously as well as those of methoxychlor (*37*), the trichloro-

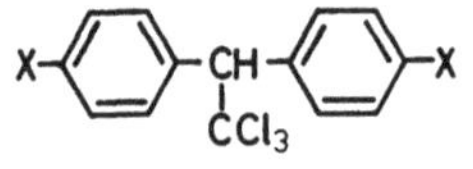

34

methyl moiety of which is replaced with various substituents, we have studied the structure-activity relationships [79]. QSAR analyses were performed for the activity in terms of the minimum molar concentration (MEC) required to induce repetitive discharges against excised nerve cord of American cockroach determined in a manner similar to that of benzyl chrysanthemates.

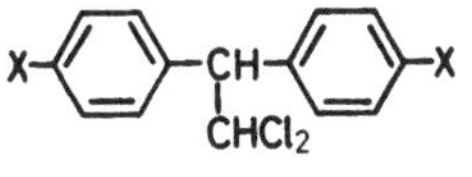

35

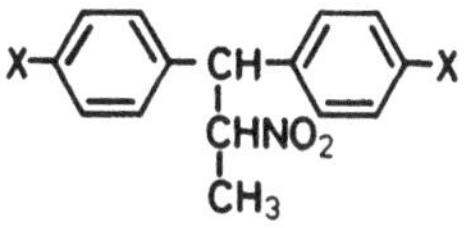

36

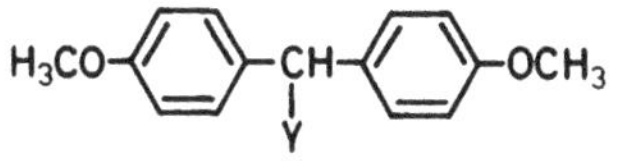

37

For aromatic substituent effects of DDT, DDD and prolan analogs, Eq. 58 was formulated,

$$\begin{aligned} \log 1/\mathrm{MEC} &= \underset{(0.328)}{4.855} + \underset{(0.395)}{2.590\Delta V_w} - \underset{(0.112)}{0.594\Delta V_w^2} + \underset{(0.295)}{0.476I} \\ n = 25 \quad & s = 0.300 \quad r = 0.955 \end{aligned} \qquad (58)$$

where I is an indicator variable for prolan analogs. There is practically no difference in the activity between DDT and DDD series if aromatic substituents are the same. The optimum ΔV_w value is about 1.94 which is close to those of Et and OEt. In the preceding section, the optimum ΔV_w very similar to this value was found for the p-substituent effect of benzyl (+)-trans-chrysanthemate on the same type of biological activity. The aromatic moiety of the pyrethroids and DDT-type compounds may fit into the target sites at the axonal membrane with a closely related (or a common) mechanism.

For the effect of benzylic substituents examined with the methoxychlor analogs, Eq. 59 was formulated.

$$\begin{aligned} \log 1/\mathrm{MEC} &= \underset{(3.220)}{3.464} + \underset{(1.548)}{3.119\Delta V_w} - \underset{(0.197)}{0.215\Delta V_w^2} \\ & \quad - \underset{(0.390)}{1.121L} - \underset{(0.141)}{0.510\pi^2} \\ n = 17 \quad & s = 0.407 \quad r = 0.962 \end{aligned} \qquad (59)$$

Eq. 59 reveals that the optimum bulkiness and the optimum hydrophobicity of substituents are located at about $\Delta V_w = 7.25$ and $\pi = 0$, respectively. Eq. 59 also shows that the shorter the length of benzylic substituents, the higher is the activity. Thus, thickset substituents are desirable for the high activity. Although the requirements for the optimum bulkiness as well as for the optimum length are not satisfied simultaneously, such thickset substituents as CBr_3, $CHEt(NO_2)$, and $C(Me)_2NO_2$ are in fact most favorable.

Holan proposed that the benzylic substituents of DDT analogs would fit into the channel of a pore in the cell membrane to induce leakage of Na^+ ion [80]. He demonstrated that the optimum dimension in diameter of the benzylic substituents is about 6–6.3 Å to plug the pore. Assuming the substituent with 6–6.3 Å diameter as a sphere, the volume is calculated as being about 78 cm^3/mol which is very close to the optimum van der Waals volume estimated from Eq. 59, which is about 76 ($= [\Delta V_w + V_w(H)] \times 10$).

5.2.4 Flower-inducing Activity of Substituted Benzoic Acids

Salicyclic acid induces flowering in some members of the Lemnaceae [81]. The chelating ability of salicylic acid had been considered to cause the flower-inducing activity. Unsubstituted benzoic acid is, however, also capable of inducing flowering [82]. Thus, we have examined the effect of ring substituents of benzoic acid on the flowering activity, quantitatively [83]. Eq. 60 was formulated for the flower inducing activity of substituted benzoic acids without chelating ability.

$$\log(1/C) = \underset{(0.62)}{4.07\sigma^{\circ}} - \underset{(0.32)}{0.90\Delta V_w(m, p)} - \underset{(0.31)}{0.52\Delta V_w(o)} + \underset{(0.37)}{1.22} \quad (60)$$
$$n = 25 \quad s = 0.361 \quad r = 0.952$$

The test plant was *Lemna paucicostata*, strain 151, grown in $^1/_{10}$ strength M medium containing 1% sucrose. C in Eq. 60 is the concentration of the unionized neutral form at pH 4.8, which is the averaged pH of the M medium during the test period of 7 days, required to flower 10% of the total plants. The steric effect in terms of van der Waals volume dependent on the substituent positions is significant in determining the activity. The hydroxyl group, when located at the ortho position, enhances the activity about 100 times those that estimated by Eq. 60, whereas reduces to $^1/_{40}$ if present at meta and para positions. Thus, the flower-inducing activity of benzoic acids is supposedly determined by steric volume and electron withdrawing effects of substituents as well as the chelating ability. For salicylic acid, the electron donating and steric effects of o-hydroxyl group counterbalance its chelating ability leading to a flower-inducing activity which does not differ much from that of unsubstituted benzoic acid.

6 Concluding Remarks

As exemplified above, various steric parameters can be properly utilized in QSAR studies depending upon the situation involved in the interaction between bioactive compounds and biomacromolecule.

In most correlations in this review, the steric parameters are supposed to be best among others. In deriving Eq. 43 for N-phenyl tetrahydrophthalimide herbicides, STERIMOL parameters are the best ones [42]. In Eq. 53 for lindane analogs, V_w works much better than MR [70]. However, this does not necessarily mean that the correlations mentioned here are always finalized. Further improvement may be possible using other steric parameters.

Since the parameter E_s, defined and estimated by Eq. 3 and 4 is based on an effective radius of substituents which may exhibit the minimum steric effect, it is expected to relate with B_1. The E_s^c value can be separated into E_s and n_H. The MR value is, in a way, a measure of the molar volume which is synonymous with V_w as described previously. Thus, certain degrees of collinearity are inevitable between partners of these pairs of related steric parameters. In some cases, the fact that the related parameter works better than the originally adopted one was recognized afterwards. For instance, Eq. 29 and 30 for enzyme reactions were originally formulated using E_s values [2,84]. With E_s^c, the quality of the correlation was much improved [25]. Verloop et al. indicated that Eq. 19 for the N-phenyl cyclic imide fungicides can be improved with STERIMOL parameters. They showed that, using B_5 parameter in place of E_s value for o-, m- and p-substituents, the HB term in Eq. 19 is unnecessary [85]. Despite these "improvements", the discussions on the structure-activity relationships are not modified very much. The collinearity between steric parameters is not a serious defect.

More important is to avoid a collinearity with the hydrophobicity parameter which could be quite high when the selection of substituents is inadequate. For this purpose, we have quite often put such substituents as $-SO_2CH_3$ and $-SO_2N(CH_3)_2$ into the correlation because of their high bulkiness as well as their low hydrophobicity.

In order to initiate the mechanism leading to an eventual effect, biologically active compounds must interact with specific groups of a certain biomacromolecular target located with a particular spatial arrangement. When the specific groups are located within a narrow cleft on a topological architecture of macromolecules, the steric fit for bioactive compounds could be achieved by being engulfed into the cleft with proper orientation and conformation. Sometimes, a conformational change of the macromolecule would be required for the proper steric fit to create a cavity. When the interaction occurs on a broad cleft (or surface) of the macromolecules, the steric effect could be less specific in fitting such target sites.

As proposed earlier in this article, the V_w value is the parameter for volume-dependent steric effects such as those on the "cavity" formation. The E_s, E_s^c and STERIMOL values are those for the width of substituents. They are supposedly applicable to cases occurring on the broad cleft. The effect on the engulfment within the narrower cleft could be rationalized by either of the volume or the width parameters depending upon the situation.

The fact that a variety of biological activities exerted by a wide range of compounds was nicely analyzable using steric parameters indicates that the participation of various types of steric effect is really plausible. But, no one could observe what is really occuring at the target site. Thus, discussions relating to the target level interactions are admittedly quite speculative. In this respect, a procedure developed quite recently by Hansch et al. [86] is noteworthy. They attempted to combine the QSAR methodology with the X-ray crystallography as well as with the computer graphics for some series of enzyme-substrate intermediate complex. With the aid of visualized three-dimensional images not only of enzyme-substrate but also of drug-receptor complexes, the nature of the interaction between partners can be identified on the atomic level. The correspondence with factors deduced from the QSAR correlation was shown to be quite reasonable. Thus, if this procedure is developed further, it will be a powerful tool to give a concrete background to QSAR correlations including the real mechanism of steric interactions.

7 References

1. Hansch, C. et al.: Nature *194*, 178 (1962); Hansch, C., Fujita, T.: J. Amer. Chem. Soc. *86*, 1616 (1964)
2. Hansch, C. et al.: ibid. *87*, 2738 (1965)
3. Hansch, C., Calef, D. F.: J. Med. Chem. *41*, 1240 (1976)
4. Taft, R. W.: Separation of Polar, Steric and Resonance Effects in Reactivity, in: Steric Effects in Organic Chemistry (ed. Newman, M. S.) John Wiley, New York 1956, p. 556
5. Charton, M.: J. Amer. Chem. Soc. *91*, 615 (1969)
6. Kutter, E., Hansch, C.: J. Med. Chem. *12*, 647 (1969)
7. Fujita, T., Nishioka, T.: Prog. Phys. Org. Chem. *12*, 49 (1976)
8. Fujita, T.: Anal. Chim. Acta *133*, 667 (1981)
9. Swain, C. G., Lupton, E. C.: J. Amer. Chem. Soc. *90*, 4328 (1968)
10. Hansch, C. et al.: J. Med. Chem. *16*, 1207 (1973)
11. Fujita, T.: ibid. *16*, 923 (1973)
12. Neims, A. H.: Biochemistry *5*, 203 (1966)
13. Mazrimas, J. A. et al.: Arch. Biochem. Biophys. *100*, 409 (1963)
14. Nishioka, T. et al.: Pestic. Biochem. Physiol. *7*, 107 (1977)
15. Nishioka, T. et al.: J. Org. Chem. *40*, 2520 (1975)
16. Fujita, T. et al.: J. Med. Chem. *20*, 1071 (1977)
17. Fujinami, A. et al.: Pestic. Biochem. Physiol. *6*, 287 (1976)
18. Takayama, C., Fujinami, A.: ibid. *12*, 163 (1979)
19. Hancock, K. et al.: J. Amer. Chem. Soc. *83*, 4211 (1961)
20. Fujita, T. et al.: J. Org. Chem. *38*, 1623 (1973)
21. MacPhee, J. A. et al.: ibid. *45*, 1164 (1980)
22. Takayama, C. et al.: ibid. *44*, 2871 (1979)
23. Takayama, C., Fujita, T.: Unpublished
24. Hansch, C., Lien, E. J.: Biochem. Pharmacol. *17*, 709 (1968)
25. Hansch, C.: Drug Design *1*, 271 (1971)
26. Eto, M. et al.: Agric. Biol. Chem. *40*, 2113 (1976); Fujita, T.: Pure Appl. Chem. *50*, 987 (1978)
27. Bellet, E. M., Casida, J. E.: Science *182*, 1135 (1973)
28. Petajan, J. H. et al.: ibid. *187*, 742 (1975)
29. Bowery, N. G. et al.: Nature *261*, 601 (1976)
30. Kirino, O. et al.: Agric. Biol. Chem. *44*, 31 (1980)
31. Kirino, O. et al.: ibid. *45*, 2669 (1981); J. Pestic. Sci. in press
32. Wilson, I. B., Cabib, E.: J. Amer. Chem. Soc. *78*, 202 (1956)
33. Verloop, A. et al.: Drug Design *7*, 165 (1976)
34. Verloop, A. et al.: Applications of STERIMOL Parameters to Drug Design, in: Steric Effects in Drug Design (eds. Motoc, I., Charton, M.) Springer, Berlin, Heidelberg, New York, 1983, p. 137
35. Nishioka, T., Fujita, T.: Unpublished
36. Van Etten, R. L. et al.: J. Amer. Chem. Soc. *89*, 3242 (1967)
37. Cohen, L. A., Takahashi, S.: ibid. *95*, 443 (1973)
38. White, G. A., Thorn, G. D.: Pestic. Biochem. Physiol. *5*, 380 (1975)
39. Voleani, B. E. et al.: J. Biol. Chem. *207*, 411 (1954)
40. Heindel, N. D. et al.: J. Med. Chem. *11*, 369 (1968)
41. Kirino, O. et al.: Agric. Biol. Chem. *44*, 2143 (1980)
42. Ohta, H. et al.: Pestic. Biochem. Physiol. *14*, 153 (1980)
43. Skoog, F. et al.: Phytochem. *6*, 1169 (1967)
44. Iwamura, H. et al.: ibid. *19*, 1309 (1980)
45. Shantz, E. M., Steward, F. C.: J. Amer. Chem. Soc. *77*, 6351 (1955)
46. Bruce, M. I., Zwar, J. A.: Proc. Roy. Soc. *B165*, 245 (1966)
47. Iwamura, H. et al.: Phytochem. *18*, 217 (1979)
48. Iwamura, H. et al.: ibid. *18*, 1265 (1979)
49. Furukawa, S., Tomizawa, Z.: Kogyokagaku Zasshi *23*, 342 (1920)

50. Acton, E. M. et al.: J. Agr. Food Chem. *18*, 1061 (1970)
51. Acton, E. M., Stone, H.: Science *193*, 584 (1976)
52. Iwamura, H.: J. Med. Chem. *23*, 308 (1980)
53. Mazur, R. H. et al.: J. Amer. Chem. Soc. *91*, 2684 (1969)
54. Mazur, R. H. et al.: J. Med. Chem. *13*, 1217 (1970)
55. Mazur, R. H. et al.: ibid. *16*, 1284 (1973)
56. Ariyoshi, Y. et al.: Bull. Chem. Soc. Japan *47*, 326 (1974)
57. Ariyoshi, Y.: Agric. Biol. Chem. *40*, 983 (1976)
58. Miyoshi, M. et al.: Bull. Chem. Soc. Japan *51*, 1433 (1978)
59. Fujino, M. et al.: Chem. Pharm. Bull. *24*, 2112 (1976)
60. Brussel, L. B. P., et al.: Z. Lebensm. Unter-Forsch. *159*, 337 (1975)
61. Iwamura, H.: J. Med. Chem. *24*, 572 (1981)
62. Pearlman, R. S.: Molecular Surface Areas and Volumes and Their Use in Structure/Activity Relationships, in: Physical Chemical Properties of Drugs (eds. Yalkowsky, S. H. et al.) Marcel Dekker, New York, 1980, p. 321
63. Lien, E. J., Gudauskas, G. A.: J. Pharm. Sci. *62*, 1968 (1973)
64. Bird, A. E.: ibid. *64*, 1671 (1975)
65. Charton, M., Charton, B. I.: J. Org. Chem. *44*, 2284 (1979)
66. Leo, A. et al.: J. Med. Chem. *19*, 611 (1976)
67. Moriguchi, I. et al.: Chem. Pharm. Bull. *24*, 1799 (1976)
68. Hopfinger, A. J.: J. Amer. Chem. Soc. *120*, 7196 (1980)
69. Battershell, C. et al.: J. Med. Chem. *24*, 812 (1981)
70. Hopfinger, A. J.: ibid. *24*, 818 (1981)
71. Hopfinger, A. J., Potenzone, R.: Mol. Pharmacol. *21*, 187 (1982)
72. Bondi, A.: J. Phys. Chem. *68*, 441 (1964)
73. Uchida, M. et al.: Pestic. Biochem. Physiol. *5*, 253 (1975)
74. Kiso, M. et al.: ibid. *8*, 33 (1978)
75. Narahashi, T.: Adv. Insect Physiol. *8*, 1 (1971)
76. Nakagawa, S. et al.: Pestic. Biochem. Physiol. *17*, 243 (1982)
77. Nakagawa, S. et al.: ibid. *17*, 259 (1982)
78. Elliott, M.: Bull. WHO *44*, 315 (1970)
79. Nishimura, K., Fujita, T.: J. Pestic. Sci. *8*, 69 (1983)
80. Holan, G.: Environ. Qual. Safety Suppl. *3*, 360 (1975)
81. Cleland, C. F., Ajami, A.: Plant Physiol. *54*, 904 (1974)
82. Watanabe, K., Takimoto, A.: Plant Cell Physiol. *20*, 847 (1979)
83. Watanabe, K. et al.: ibid. *22*, 1469 (1981)
84. Hansch, C., Deutsch, E. W.: Biochim. Biophys. Acta *126*, 117 (1966)
85. Verloop, A.: The STERIMOL Approach, Further Development of the Method and New Applications, in: Proc. of 5th Int. Congr. of Pesticide Chemistry Vol. 1 (eds. Miyamoto, J. et al.) Pergamon, Oxford 1983, in press
86. Hansch, C. et al.: J. Med. Chem. *25*, 777 (1982)

Bei Fragen zur Produktsicherheit wenden Sie sich bitte an:
If you have any questions regarding product safety,
please contact:

Walter de Gruyter GmbH
Genthiner Straße 13
10785 Berlin
productsafety@degruyterbrill.com